T0225085

## Algebraic Number Theory for Beginners

This book introduces algebraic number theory through the problem of generalizing "unique prime factorization" from ordinary integers to more general domains. Solving polynomial equations in integers leads naturally to these domains, but unique prime factorization may be lost in the process. To restore it, we need Dedekind's concept of *ideals*. However, one still needs the supporting concepts of algebraic number field and algebraic integer, and the supporting theory of rings, vector spaces, and modules. It was left to Emmy Noether to encapsulate the properties of rings that make unique prime factorization possible, in what we now call *Dedekind rings*. The book develops the theory of these concepts, following their history, motivating each conceptual step by pointing to its origins, and focusing on the goal of unique prime factorization with a minimum of distraction or prerequisites. This makes for a self-contained, easy-to-read book, short enough for a one-semester course.

JOHN STILLWELL is the author of many books on mathematics; among the best known are *Mathematics and Its History, Naive Lie Theory*, and *Elements of Mathematics*. He is a member of the inaugural class of Fellows of the American Mathematical Society and winner of the Chauvenet Prize for mathematical exposition.

# Algebraic Number Theory for Beginners

## Following a Path from Euclid to Noether

### JOHN STILLWELL
*University of San Francisco*

## CAMBRIDGE
### UNIVERSITY PRESS

University Printing House, Cambridge CB2 8BS, United Kingdom

One Liberty Plaza, 20th Floor, New York, NY 10006, USA

477 Williamstown Road, Port Melbourne, VIC 3207, Australia

314321, 3rd Floor, Plot 3, Splendor Forum, Jasola District Centre,
New Delhi 110025, India

103 Penang Road, #05–06/07, Visioncrest Commercial, Singapore 238467

Cambridge University Press is part of the University of Cambridge.

It furthers the University's mission by disseminating knowledge in the pursuit of
education, learning, and research at the highest international levels of excellence.

www.cambridge.org
Information on this title: www.cambridge.org/9781316518953
DOI: 10.1017/9781009004138

First published 2022

A catalogue record for this publication is available from the British Library.

ISBN 978-1-316-51895-3 Hardback
ISBN 978-1-009-00192-2 Paperback

To my grandchildren, Ida and Isaac

# Contents

*Preface*                                                              *page* xi
*Acknowledgments*                                                            xiv

**1    Euclidean Arithmetic**                                                  1
1.1   Divisors and Primes                                                      2
1.2   The Form of the gcd                                                      5
1.3   The Prime Divisor Property                                               8
1.4   Irrational Numbers                                                      10
1.5   The Equation $x^2 - 2y^2 = 1$                                           13
1.6   Rings                                                                   15
1.7   Fields                                                                  19
1.8   Factors of Polynomials                                                  22
1.9   Discussion                                                             24

**2    Diophantine Arithmetic**                                               33
2.1   Rational versus Integer Solutions                                       34
2.2   Fermat's Last Theorem for Fourth Powers                                 36
2.3   Sums of Two Squares                                                     38
2.4   Gaussian Integers and Primes                                            41
2.5   Unique Gaussian Prime Factorization                                     43
2.6   Factorization of Sums of Two Squares                                    45
2.7   Gaussian Primes                                                         47
2.8   Primes that Are Sums of Two Squares                                     48
2.9   The Equation $y^3 = x^2 + 2$                                            50
2.10  Discussion                                                             53

**3    Quadratic Forms**                                                      59
3.1   Primes of the Form $x^2 + ky^2$                                         60
3.2   Quadratic Integers and Quadratic Forms                                  61
3.3   Quadratic Forms and Equivalence                                        63

3.4    Composition of Forms                                    66
3.5    Finite Abelian Groups                                   68
3.6    The Chinese Remainder Theorem                           71
3.7    Additive Notation for Abelian Groups                    73
3.8    Discussion                                              74

**4      Rings and Fields**                                    **78**
4.1    Integers and Fractions                                  79
4.2    Domains and Fields of Fractions                         82
4.3    Polynomial Rings                                        83
4.4    Algebraic Number Fields                                 86
4.5    Field Extensions                                        89
4.6    The Integers of an Algebraic Number Field              93
4.7    An Equivalent Definition of Algebraic Integer           96
4.8    Discussion                                              99

**5      Ideals**                                              **104**
5.1    "Ideal Numbers"                                         105
5.2    Ideals                                                  108
5.3    Quotients and Homomorphisms                             111
5.4    Noetherian Rings                                        113
5.5    Noether and the Ascending Chain Condition               116
5.6    Countable Sets                                          119
5.7    Discussion                                              121

**6      Vector Spaces**                                       **126**
6.1    Vector Space Basis and Dimension                        127
6.2    Finite-Dimensional Vector Spaces                        130
6.3    Linear Maps                                             134
6.4    Algebraic Numbers as Matrices                           136
6.5    The Theorem of the Primitive Element                    139
6.6    Algebraic Number Fields and Embeddings in $\mathbb{C}$  142
6.7    Discussion                                              144

**7      Determinant Theory**                                  **149**
7.1    Axioms for the Determinant                              150
7.2    Existence of the Determinant Function                   153
7.3    Determinants and Linear Equations                       156
7.4    Basis Independence                                      159
7.5    Trace and Norm of an Algebraic Number                   161
7.6    Discriminant                                            164
7.7    Discussion                                              168

| **8** | **Modules** | 171 |
| --- | --- | --- |
| 8.1 | From Vector Spaces to Modules | 172 |
| 8.2 | Algebraic Number Fields and Their Integers | 174 |
| 8.3 | Integral Bases | 176 |
| 8.4 | Bases and Free Modules | 179 |
| 8.5 | Integers over a Ring | 182 |
| 8.6 | Integral Closure | 184 |
| 8.7 | Discussion | 186 |
| **9** | **Ideals and Prime Factorization** | 189 |
| 9.1 | To Divide Is to Contain | 190 |
| 9.2 | Prime Ideals | 192 |
| 9.3 | Products of Ideals | 194 |
| 9.4 | Prime Ideals in Algebraic Number Rings | 196 |
| 9.5 | Fractional Ideals | 197 |
| 9.6 | Prime Ideal Factorization | 199 |
| 9.7 | Invertibility and the Dedekind Property | 201 |
| 9.8 | Discussion | 204 |
| | *References* | 211 |
| | *Index* | 217 |

# Preface

The history of mathematics, like the life of each individual mathematician, is a story that begins with concrete experience and (generally) ends at high levels of abstraction. A good example, which we follow in this book, is the story of arithmetic. It begins with *counting*, then *adding* and *multiplying*; then it symbolizes this experience in *equations*. Next, it investigates equations via the abstract structures of *groups*, *rings*, and *fields*, and so on, to higher and higher levels of abstraction. This is a typical story, but the story alone does not explain why abstraction is necessary – or why it ever happened at all.

The reason is that abstract structures distill the essence of many concrete structures, enabling us to see past a mass of distracting details. For example, it is an impossible task to list all the facts about addition and multiplication of numbers, and some specific questions about them were not answered for hundreds of years. Mathematicians have been able to answer some of the hard questions only by working with abstract concepts that encapsulate the nature of addition and multiplication.

The art of algebra is the art of abstraction: choosing concepts that distill the essence of questions that interest us. To some extent the proof that we have chosen the "right" concepts is in the pudding. The right concepts answer many questions and make the answers seem obvious. But a concept may be "right" in the sharper sense that we can prove it is a *necessary part of the answer*. That is, an answer or solution exists only in structures that exemplify the concept in question.

A famous example is the discovery by Galois of the group concept, which explains which polynomial equations have solutions by radicals. Galois associated a group – now called the *Galois group* – with each equation and showed that an equation is solvable by radicals if and only if its Galois group

xi

has a certain property, now called *solvability*. Thus the concept of solvable group is the "right" concept to explain solvability of equations.

In this book we study a second famous example: Dedekind's theory of rings and ideals, which explains the phenomenon of unique prime factorization in arithmetic and its generalizations. Again, there is an abstract algebraic concept – now called a *Dedekind domain* – that exactly captures the property of unique prime factorization. Dedekind domains are an equally good example of the power of abstraction, and in some ways easier than the group concept, since their algebra is *commutative*. They also have a natural motivation as an outgrowth of arithmetic – which is why our path starts with Euclid.

The material in the book may be found in comprehensive graduate algebra texts, such as Zariski and Samuel (1958), Jacobson (1985), and Rotman (2015), but it is hard work to extract it from them. I prefer not to be comprehensive, so as to tell the story with only the essential abstractions, and to make it sufficiently self-contained to be accessible to undergraduates. This means including enough number theory to motivate the problem of unique prime factorization, which we do in the first three chapters. These chapters introduce algebraic numbers to solve classical equations such as the Pell equation, and the concepts of ring and field that abstract the algebra *of* these numbers.

Accessibility to undergraduates, in my opinion, also means including the *linear algebra* needed to view number fields and number rings as vector spaces and modules. I realize that this opinion is somewhat controversial. Modern books on algebraic number theory commonly assume linear algebra is already known, and indeed, every undergraduate takes a course in linear algebra these days. But linear algebra is a multifaceted subject, and I doubt that many undergraduates know the subject from the viewpoint needed here, which varies the base field (or base *ring*) and relies on the trace, determinant, characteristic polynomial, and discriminant. Those who do may skip the parts where these topics are covered, but I believe they should at least be skimmed in order to see where linear algebra fits in the bigger algebraic picture.

In fact the book closest to this one could be the classic telling of the story by Dedekind in 1877, which may be seen in English translation in Dedekind (1996). Dedekind's account is at a lower level of generality than ours, being concerned only with the needs of number theory, but it follows a similar path. The advantage of raising the level of generality is that one sees how close Dedekind came to the ultimate setting for unique prime factorization. As Emmy Noether used to say: "Es steht alles schon bei Dedekind." ("Everything is already in Dedekind.")

I should say, however, that I raise the level of generality only in easy stages, when it becomes necessary. As in the history of the subject, the general case appears only after the important special cases.

To make the book useful to undergraduates and instructors, I have included many exercises, distributed in small batches at the end of most sections. These range from routine exercises, which test and reinforce understanding of new concepts, to exercise "packages" leading to substantial theorems. These theorems are often concrete consequences of the abstract machinery developed in the main text. The aim of each "package" is to reach an interesting goal by a sequence of easy steps, so the exercises include commentary to explain what the goal is and (in some cases) where to look for help later in the book.

Although many important and useful results occur in exercises, it should be stressed that these results are *not assumed* in the main text. In a few cases they are later *used* in the main text, but only after the main text has proved them.

In fact, the technical prerequisites for this book are small, since the whole point is to grow a big abstract structure from ideas in arithmetic. High school algebra should suffice, if it includes the matrix concept, and otherwise undergraduate linear algebra as far as matrices. Apart from these technical skills, however, the reader will also need sufficient mathematical maturity to be comfortable with abstractions. In most cases this will mean a couple of years of undergraduate mathematics, even a first course in abstract algebra. This book carries commutative algebra far beyond the typical first course, but it certainly will not hurt to have a first impression of fields, rings, and ideals.

# Acknowledgments

As usual, my greatest thanks go to my wife Elaine, who did the first round of proofreading and picked up many errors. Others were found by Mark Hunacek, Paul Stanford, and an anonymous reviewer, who also made valuable suggestions that clarified several points. I also thank the University of San Francisco and the DPMMS at the University of Cambridge for support during the writing of the book.

# 1

# Euclidean Arithmetic

## Preview

Euclid's *Elements*, from around 300 BCE, is the source of many basic parts of modern mathematics, such as geometry, the axiomatic method, and the theory of real numbers. It is also the source of *arithmetic* as mathematicians know it: the theory of addition and multiplication of natural numbers, with emphasis on the concepts of divisibility and primes.

For Euclid, a natural number $b$ is a **divisor** of a natural number $a$ if

$$a = bc \quad \text{for some natural number } c.$$

Then a natural number $p > 1$ is *prime* if its only divisors are itself and 1. These concepts lead, as Euclid showed by a short but ingenious proof, to the discovery that there are infinitely many primes.

Even more ingeniously, Euclid proved the **prime divisor property**: If a prime $p$ divides a product $ab$, then $p$ divides $a$ or $p$ divides $b$. His proof is based on the famous **Euclidean algorithm** for finding the greatest common divisor of two natural numbers. The prime divisor property easily implies what we now call the **fundamental theorem of arithmetic**, or **unique prime factorization**: Every natural number greater than 1 may be expressed uniquely (up to the order of factors) as a product of primes.

Unique prime factorization is so useful that mathematicians would like it to hold wherever the concept of "factorization" makes sense. In fact, as we will see in later chapters, even when it is lost they will try to recover it. In this chapter we prepare to explore more general domains for factorization by introducing the concepts (and some examples) of **ring** and **field**.

## 1.1 Divisors and Primes

In this chapter we will be working mainly with the set $\mathbb{N} = \{0, 1, 2, 3, 4, 5, \ldots\}$ of **natural numbers**. These are the numbers obtained from 0 by "counting": that is, by repeatedly adding 1. It follows (informally) that from any natural number $n$ we can reach 0 in a finite number of steps by "counting backwards," and hence that *any set of natural numbers has a least member*. Since Euclid, this so-called **well-ordering** property of $\mathbb{N}$ has been the basis of virtually all reasoning about the natural numbers, so it is usually taken as an axiom. In this section we will use it, as Euclid did, to prove results about divisibility and primes.

We have already said what it means for a natural number $b$ to divide a natural number $a$; namely, $a = bc$ for some natural number $c$. So if $b$ does *not* divide $a$, we necessarily have, for any natural number $q$,

$$a = bq + r, \quad \text{with } r > 0.$$

When $r$ is least possible, we call $q$ the **quotient** (of $a$ by $b$) and $r$ the **remainder**. It then follows that $0 < r < b$, because if $r = b + r'$, we would have

$$a = b(q + 1) + r', \quad \text{contrary to the assumption that } r \text{ is the least remainder.}$$

The two cases, where $b$ does and does not divide $a$, can be combined in the following **division property**: For any natural numbers, there are natural numbers $q$ and $r$ such that

$$a = bq + r, \quad \text{where } 0 \leq r < b. \tag{*}$$

This property is often misleadingly called the "division algorithm." (It is not an algorithm, but it paves the way for the very important Euclidean algorithm, as we will see in the next section.) Finding the quotient and remainder for a given pair $a, b$ is called **division with remainder**.

Another easy application of well-ordering of $\mathbb{N}$ tells us that *every natural number greater than 1 is divisible by a prime*. Start with any natural number $a > 1$. If $a$ is not prime, then $a = bc$ for some smaller numbers $b$ and $c$. Then if $b$ is not prime, we have $b = de$ for some smaller natural numbers $d$ and $e$, and so on. Since natural numbers cannot decrease forever, this process must halt – necessarily with a prime $p$ that divides $a$. It follows, by repeatedly finding prime divisors, that *every natural number has a prime factorization*.

With these easy properties of divisors and primes, we are now ready for something ingenious: Euclid's proof that there are infinitely many primes.

**Infinitude of primes.** *For any prime numbers* $p_1, p_2, \ldots, p_k$, *there is a prime number* $p_{k+1} \neq p_1, p_2, \ldots, p_k$.

*Proof.* Consider the number $N = (p_1 \cdot p_2 \cdots p_k) + 1$. None of $p_1, p_2, \ldots, p_k$ divide $N$ because they each leave remainder 1. But *some* prime divides $N$ because $N > 1$. This prime is the $p_{k+1}$ we seek.     □

The beauty of this proof is that it avoids having to find any pattern in the sequence of primes, or finding divisors of a number, both of which are hard problems.

### 1.1.1 The Euclidean Algorithm

Although it is hard to find the divisors of a given (large) natural number, it is surprisingly quick and easy to find *common* divisors of two natural numbers. This can be done by the **Euclidean algorithm** for finding the **greatest common divisor** $\gcd(a, b)$ of two natural numbers $a$ and $b$. As Euclid described it, (*Elements*, Book VII, Proposition 1) the algorithm "repeatedly subtracts the lesser number from the greater." More formally, it repeatedly replaces the pair $\{a, b\}$, where $a > b$, by the pair $\{b, a - b\}$ until the members of the pair become equal – at which stage each member is $\gcd(a, b)$.

For example, if we begin with the pair $\{34, 21\}$, the pairs produced by the algorithm are the following

$$\{34, 21\} \to \{21, 13\} \to \{13, 8\} \to \{8, 5\} \to \{5, 3\} \to \{3, 2\} \to \{2, 1\} \to \{1, 1\}.$$

And we conclude that $\gcd(34, 21) = 1$.

In general, the correctness of the Euclidean algorithm is guaranteed by the following theorem.

**Euclidean algorithm produces the gcd.** *If the Euclidean algorithm is applied to two natural numbers* $a, b > 0$, *then it terminates in a finite number of steps with the pair whose members are both* $\gcd(a, b)$.

*Proof.* Suppose that $d$ is any common divisor of $a$ and $b$, where $a > b$. This means that $a = a'd$ and $b = b'd$ for some $a', b' > 0$, and hence that

$$a - b = (a' - b')d.$$

Thus, $d$ is also a divisor of $a - b$. There is a similar proof that any common divisor of two numbers is also a divisor of their sum, so a divisor of $b$ and $a - b$ is also a divisor of $b + (a - b) = a$. It follows that *each* pair produced by the Euclidean algorithm has the same common divisors, and hence the same gcd.

Now, as long as the pairs produced by the algorithm are unequal, subtraction occurs, and it will decrease the sum of the two members of the pair. By the well-ordering of $\mathbb{N}$, the sum cannot decrease forever, so the algorithm necessarily halts with a pair of equal numbers. Being equal, they equal their own gcd; hence they each equal $\gcd(a, b)$.                                                                  □

In practice it is usual to speed up the Euclidean algorithm by doing **division with remainder** instead of subtraction. That is, we replace the pair $\{a, b\}$, where $a > b$, with the pair $\{b, r\}$, where $r$ is the remainder when $a$ is divided by $b$. This process is simply a shortening of repeated subtraction, because $r$ can be found by subtracting $b$ repeatedly from $a$. However, the usual "long division" process generally finds $r$ more quickly than repeated subtraction.

In fact, by using division with remainder, we can be sure that the number of steps required for the Euclidean algorithm to halt is roughly proportional to the number of decimal digits in $a$. The example above, incidentally, is one where each division with remainder is actually the same as a single subtraction. This happens whenever $a$ and $b$ are a pair of consecutive **Fibonacci numbers**: the numbers $0, 1, 1, 2, 3, 5, 8, 13, 21, 34, 55, 89, 144, 233, \ldots$ defined by

$$F_0 = 0, \quad F_{n+2} = F_{n+1} + F_n.$$

This is the case where the Euclidean algorithm runs most slowly. But even here, the number of steps is roughly proportional to the number of decimal digits.

## Exercises

1. Explain why the Euclidean algorithm, applied to the pair $\{F_{n+2}, F_{n+1}\}$, yields all preceding pairs of consecutive Fibonacci numbers.
2. Deduce that $\gcd(F_{n+2}, F_{n+1}) = 1$.

Division with remainder is the preferred way to run the Euclidean algorithm in practice, because it is generally faster. But it also has advantages in theory, since it applies in situations (such as division of polynomials) where division with remainder is *not* achievable by repeated subtraction. In the case of ordinary positive integers $a, b$, the process of repeated division with remainder can be elegantly "frozen in time" by the so-called **continued fraction** for $a/b$.

Given positive integers $a > b$, the continued fraction process finds $q_1 > 0$ and $r_1 \geq 0$ ("quotient" and "remainder") such that $a = bq_1 + r_1$ with $r_1 < b_1$, and we write down the equivalent equation

$$\frac{a}{b} = q_1 + \frac{r_1}{b}.$$

If $r_1 = 0$, then the process ends there, because we have found that $b$ divides $a$ and hence that $\gcd(a, b) = b$.

If $r_1 > 0$, then we rewrite the above equation as

$$\frac{a}{b} = q_1 + \frac{1}{b/r_1}$$

and repeat the process on the fraction $b/r_1$ (which we can do since $b > r_1 > 0$). In this way we can simulate the action of the Euclidean algorithm on a pair $(a, b)$ by the process of "continuing" a fraction $a/b$.

3. Explain why the continued fraction process terminates for any positive integers $a, b$.
4. Applying the continued fraction process to 23 and 5, show that

$$\frac{23}{5} = 4 + \cfrac{1}{1 + \cfrac{1}{1 + \cfrac{1}{2}}}$$

Division with remainder also has a neat representation by $2 \times 2$ matrices, in which division with remainder corresponds to *extracting a matrix factor* from a column vector. In this setup, the pair $\{a, b\}$ is represented by the column vector

$$\begin{pmatrix} a \\ b \end{pmatrix}, \quad \text{where } a > b.$$

5. If $a = q_1 b + r_1$, show that $\begin{pmatrix} a \\ b \end{pmatrix} = \begin{pmatrix} q_1 & 1 \\ 1 & 0 \end{pmatrix} \begin{pmatrix} b \\ r_1 \end{pmatrix}$.

Then, if $b > r_1 \neq 0$, one can repeat the process on the column vector $\begin{pmatrix} b \\ r_1 \end{pmatrix}$.

6. Show in particular that $\begin{pmatrix} 23 \\ 5 \end{pmatrix} = \begin{pmatrix} 4 & 1 \\ 1 & 0 \end{pmatrix} \begin{pmatrix} 1 & 1 \\ 1 & 0 \end{pmatrix} \begin{pmatrix} 1 & 1 \\ 1 & 0 \end{pmatrix} \begin{pmatrix} 2 & 1 \\ 1 & 0 \end{pmatrix} \begin{pmatrix} 1 \\ 0 \end{pmatrix}$.

## 1.2 The Form of the gcd

The correctness of the Euclidean algorithm says that $\gcd(a, b)$ results from the pair $\{a, b\}$ by repeated subtraction. This implies that $\gcd(a, b)$ has a very simple symbolic form. Because subtraction is involved, the form involves **integers**; that is, natural numbers and their negatives. The system of integers is denoted by $\mathbb{Z}$, from the German word "Zahlen" for numbers.

**Form of the gcd.** *For any natural numbers $a, b > 0$, there are $m, n \in \mathbb{Z}$ such that*

$$\gcd(a, b) = ma + nb.$$

*Proof.* We show in fact that the numbers produced from $a, b$ at *each step* of the Euclidean algorithm are of the form $ma + nb$. This is certainly true at the beginning, where $a = 1 \cdot a + 0 \cdot b$ and $b = 0 \cdot a + 1 \cdot b$.

And if the pair at some stage is $\{m_1 a + n_1 b, m_2 a + n_2 b\}$, then the pair at the next stage is $\{m_2 a + n_2 b, (m_1 - m_2)a + (n_1 - n_2)b\}$, which again consists of numbers of the required form.

Thus, the numbers at all stages are of the form $ma + nb$. In particular, this is true at the last stage, when each number is $\gcd(a, b)$.      □

Given a pair of moderately sized numbers $a, b$ (say, two-digit numbers), it may be hard to spot $m$ and $n$ such that $\gcd(a, b) = ma + nb$. However, $m$ and $n$ are easily computed by running the Euclidean algorithm on the letters $a$ and $b$, doing exactly the same subtractions on the symbolic forms that we originally did on numbers. For example, here is what happens when we run the numerical and symbolic computations side by side in the case where $a = 34$ and $b = 21$.

$$
\begin{aligned}
&\{34, 21\} & &\{a, b\} \\
\to\ &\{21, 34 - 21\} = \{21, 13\} &\to\ &\{b, a - b\} \\
\to\ &\{13, 21 - 13\} = \{13, 8\} &\to\ &\{a - b, b - (a - b)\} = \{a - b, -a + 2b\} \\
\to\ &\{8, 13 - 8\} = \{8, 5\} &\to\ &\{-a + 2b, a - b - (-a + 2b)\} = \{-a + 2b, 2a - 3b\} \\
\to\ &\{5, 8 - 5\} = \{5, 3\} &\to\ &\{2a - 3b, -a + 2b - (2a - 3b)\} = \{2a - 3b, -3a + 5b\} \\
\to\ &\{3, 5 - 3\} = \{3, 2\} &\to\ &\{-3a + 5b, 2a - 3b - (-3a + 5b)\} = \{-3a + 5b, 5a - 8b\} \\
\to\ &\{2, 3 - 2\} = \{2, 1\} &\to\ &\{5a - 8b, -3a + 5b - (5a - 8b)\} = \{5a - 8b, -8a + 13b\}.
\end{aligned}
$$

From the last line we read off $1 = \gcd(a, b) = -8a + 13b$, and it can be checked that indeed $1 = -8 \cdot 34 + 13 \cdot 21$.

The symbolic form of the Euclidean algorithm, and hence of the gcd, was not known to Euclid. Indeed, written calculation with numbers did not develop until centuries after him, because numerical calculation could be done perfectly well with the abacus. And it was not until the sixteenth century that mathematicians realized that written calculation with symbols ("algebra") was a powerful idea – in fact more powerful than written calculation with numbers. Still, even with the primitive notation at his disposal, Euclid was able to prove the **prime divisor property**, the main result of the next section.

### 1.2.1 Linear Diophantine Equations

The equation $ax + by = c$, where $a, b, c$ are integers, becomes interesting when integer solutions for $x$ and $y$ are sought. The equation obviously has no such solution when $\gcd(a,b)$ does not divide $c$, because in that case $\gcd(a,b)$ divides $ax + by$ but not $c$. However, this is the only obstruction.

**Criterion for solvability.** *If* $\gcd(a,b)$ *divides* $c$, *then* $ax + by = c$ *has an integer solution.*

*Proof.* It follows from the above that $\gcd(a,b) = ma + nb$ for some integers $m$ and $n$. Then, if $c = d \cdot \gcd(a,b)$, it follows that $ax + by = c$ for $x = dm$ and $y = dn$. □

This criterion for solvability generalizes to linear equations in more than two variables. For example, $ax + by + cz = d$ has an integer solution $\Leftrightarrow$ $\gcd(a,b,c)$ divides $d$. The ($\Rightarrow$) direction is clear, for the same reason as above. The ($\Leftarrow$) direction holds because

$$\gcd(a,b,c) = la + mb + nc \quad \text{for some integers } l,m,n,$$

which follows from the above because $\gcd(a,b,c) = \gcd(\gcd(a,b),c)$.

We also know that we can find the required $m,n$ for $\gcd(a,b)$ by the extended Euclidean algorithm described above. Finally, we can find *all* solutions of $ax + by = c$ by adding to any single solution the solutions of $ax + by = 0$, which are $x = kb/\gcd(a,b)$, $y = -ka/\gcd(a,b)$ for all integers $k$.

With these observations we can move on to Diophantine equations of higher degree. We begin in Section 1.5 with a quadratic equation in two variables. Other examples, of degree 2 and 3, are discussed in the next chapter. But first, let us see what the gcd can tell us about prime numbers.

## Exercises

1. Using the symbolic Euclidean algorithm above, find integers $m,n$ such that $13m + 17n = 1$.

The matrix version of division with remainder, explored in the previous set of exercises, can be very elegantly "inverted" to give the integers $m$ and $n$ such that $\gcd(a,b) = ma + nb$. Recall that $a = q_1 b + r_1$ is represented by the matrix equation

$$\begin{pmatrix} a \\ b \end{pmatrix} = \begin{pmatrix} q_1 & 1 \\ 1 & 0 \end{pmatrix} \begin{pmatrix} b \\ r_1 \end{pmatrix}.$$

2. Show that if repeated division with remainder on the pair $a,b$ produces successive quotients $q_1, q_2, \ldots, q_n$ and $\gcd(a,b) = d$, then

$$\begin{pmatrix} a \\ b \end{pmatrix} = \begin{pmatrix} q_1 & 1 \\ 1 & 0 \end{pmatrix} \begin{pmatrix} q_2 & 1 \\ 1 & 0 \end{pmatrix} \cdots \begin{pmatrix} q_n & 1 \\ 1 & 0 \end{pmatrix} \begin{pmatrix} d \\ 0 \end{pmatrix}.$$

3. Deduce that

$$\begin{pmatrix} d \\ 0 \end{pmatrix} = \begin{pmatrix} q_n & 1 \\ 1 & 0 \end{pmatrix}^{-1} \cdots \begin{pmatrix} q_2 & 1 \\ 1 & 0 \end{pmatrix}^{-1} \begin{pmatrix} q_1 & 1 \\ 1 & 0 \end{pmatrix}^{-1} \begin{pmatrix} a \\ b \end{pmatrix},$$

and show that

$$\begin{pmatrix} q & 1 \\ 1 & 0 \end{pmatrix}^{-1} = \begin{pmatrix} 0 & 1 \\ 1 & -q \end{pmatrix}.$$

4. Deduce from exercise 6 of Section 1.1 that

$$\begin{pmatrix} 1 \\ 0 \end{pmatrix} = \begin{pmatrix} 0 & 1 \\ 1 & -2 \end{pmatrix} \begin{pmatrix} 0 & 1 \\ 1 & -1 \end{pmatrix} \begin{pmatrix} 0 & 1 \\ 1 & -1 \end{pmatrix} \begin{pmatrix} 0 & 1 \\ 1 & -4 \end{pmatrix} \begin{pmatrix} 23 \\ 5 \end{pmatrix},$$

and hence express the gcd of 23 and 5 in the form $23m + 5n$.

Another way to prove $\gcd(a,b) = ma + nb$ is by considering the smallest positive value $c$ of $ma + nb$ for $m, n \in \mathbb{Z}$. This idea will be used in Section 5.2 to prove that $\mathbb{Z}$ is a **principal ideal domain**.

5. Show that all values of $ma + nb$ are multiples of $c$ (this part uses the division property of $\mathbb{Z}$).
6. Deduce that $c$ divides $a$ and $b$, and that any divisor of $a$ and $b$ divides $c$.
7. Conclude that $c = \gcd(a,b)$.

## 1.3 The Prime Divisor Property

The relevance of the Euclidean algorithm to the theory of primes becomes clear when we consider $\gcd(a, p)$, where $p$ is prime. If $p$ does not divide $a$, then we must have $\gcd(a, p) = 1$, because the only divisors of $p$ are 1 and $p$ itself. This leads to a crucial result.

**Prime divisor property.** *If $a$ and $b$ are natural numbers and $p$ is a prime that divides $ab$, then $p$ divides $a$ or $p$ divides $b$.*

*Proof.* Suppose that $p$ does not divide $a$, so we must prove that $p$ divides $b$. First, as we have just remarked, $\gcd(a, p) = 1$. Also, as we saw in the previous section, $\gcd(a, p) = ma + np$ for some integers $m$ and $n$, so

$$1 = ma + np \quad \text{for some integers } m \text{ and } n.$$

Multiplying both sides of this equation by $b$, we get

$$b = mab + npb \quad \text{for some integers } m \text{ and } n.$$

Since $p$ divides $ab$ by hypothesis and $p$ divides $pb$, obviously, $b$ is a sum of terms divisible by $p$. Hence, $b$ itself is divisible by $p$. □

In proving this prime divisor property, Euclid came as close as he probably could (given his poor notational resources) to proving what we now call the **fundamental theorem of arithmetic**, or **unique prime factorization**. Unique prime factorization easily follows from the prime divisor property if one has notation for arbitrary products of primes.

**Unique prime factorization.** *If $p_1, p_2, \ldots, p_k$ and $q_1, q_2, \ldots, q_l$ are prime numbers such that*

$$p_1 p_2 \cdots p_k = q_1 q_2 \cdots q_l,$$

*then the same factors occur on each side, perhaps in a different order.*

*Proof.* Since $p_1$ divides the left side of the equation, it also divides the right side, hence, it divides one of the factors $q_i$ by the prime divisor property. It follows that $p_1 = q_i$, and we may cancel $p_1$ and $q_i$ from the equation. Repeating the argument with the factors that remain, we eventually find that each $p_j$ equals some $q_k$, and vice versa; so the factors on each side are exactly the same, though perhaps in a different order. □

We sometimes express this theorem by saying that factorization of a natural number greater than 1 into primes is unique "up to the order of factors." Later, we will see many other statements of unique prime factorization, and the "uniqueness" will be "up to order" and sometimes other trivial variations. For example, prime factorization of *integers* is unique not only "up to order" but also "up to sign" because, for example, $6 = 2 \cdot 3 = (-2) \cdot (-3)$.

The next section gives some applications of unique prime factorization. Due to its usefulness and simplicity, unique prime factorization has been sought in many other domains where "factorization" makes sense. In fact, a major theme of this book is the search for appropriate concepts of "prime" in domains where the obvious kind of factorization fails to be unique.

## Exercises

In school you may have used prime factorization to find the gcd ("greatest common divisor") and the lcm ("least common multiple") of given positive integers. We can justify this idea with the help of unique prime factorization.

1. Find gcd of 60 and 84 by finding the common primes in their prime factorizations.

2. Also find $\mathrm{lcm}(60,84)$.

3. Given that $p_1, \ldots, p_k$ are the primes in the factorizations of $a$ and $b$, so

$$a = p_1^{m_1} \cdots p_k^{m_k}, \quad \text{and}$$
$$b = p_1^{n_1} \cdots p_k^{n_k}, \quad \text{for some integers } m_1, n_1, \ldots, m_k, n_k \geq 0,$$

explain why

$$\gcd(a,b) = p_1^{\min(m_1,n_1)} \cdots p_k^{\min(m_k,n_k)}$$
$$\mathrm{lcm}(a,b) = p_1^{\max(m_1,n_1)} \cdots p_k^{\max(m_k,n_k)}.$$

4. Use these formulas for gcd and lcm to prove $\gcd(a,b)\mathrm{lcm}(a,b) = ab$.

Our proof of unique prime factorization in this section comes from the division property of $\mathbb{Z}$, via the prime divisor property. In the exercises to the last section we showed that the division property also implies the **principal ideal property** of $\mathbb{Z}$, according to which the numbers of the form $ma + nb$ are all multiples of a certain nonzero member $c$. We can also prove the prime divisor property from the principal ideal property, as the following exercises show.

5. Suppose that $p$ divides $ab$, but $p$ does not divide $a$. Given that the numbers of the form $mp + na$ are all multiples of some positive $c \neq 0$, show that $c = 1$.

6. Now deduce the prime divisor property.

## 1.4 Irrational Numbers

The numbers considered so far are the natural numbers and their close relatives the integers. A still larger class whose properties derive from those of the integers is the set $\mathbb{Q}$ of **rational** numbers: the ratios, or quotients, $m/n$ of integers $m, n$ with $n \neq 0$. It was once thought that *all* numbers are rational, but that hope was dashed (and serious mathematics began) when one of the followers of Pythagoras discovered that $\sqrt{2}$ is not. This discovery shocked the Pythagoreans, who sought a "rational" (number) explanation of everything, but who also knew that $\sqrt{2}$ was a fundamental quantity in geometry – the diagonal of the unit square. We give a proof of the irrationality of $\sqrt{2}$ by a method that extends to many other numbers.

**Irrationality of $\sqrt{2}$.** *For any natural numbers $m$ and $n$, $m/n \neq \sqrt{2}$.*

*Proof.* Suppose on the contrary that $\sqrt{2} = m/n$ or, equivalently, that $2n^2 = m^2$ for some natural numbers $m$ and $n \neq 0$. Consider the prime factorizations of $2n^2$ and $m^2$, which must be the same up to the order of factors.

Now the number of occurrences of the prime factor 2 in $2n^2$ must be *odd*; namely the one visible occurrence, plus twice the number of occurrences in $n$. But the number of occurrences of 2 in $m^2$ must be *even*; namely, twice the number of occurrences in $m$.

This contradicts unique prime factorization, hence, there are no natural numbers $m$ and $n$ such that $\sqrt{2} = m/n$. □

The method of comparing numbers of particular primes on both sides of a hypothetical equation extends to prove irrationality of many square roots – in fact, it applies to the square root of any natural number that is not a perfect square. For example, $\sqrt{6} \neq m/n$ because the equation $6n^2 = m^2$, or $2 \cdot 3n^2 = m^2$, is impossible—there are an odd number of factors 2 on the left side, but an even number on the right. Once again, this contradicts unique prime factorization.

The method also extends to cube roots, fourth roots, and so on. For example, $\sqrt[3]{2} \neq m/n$ because the equivalent equation $2n^3 = m^3$ contradicts unique prime factorization. The number of factors of 2 in $m^3$ is a multiple of 3 (three times the number of 2s in the prime factorization of $m$), whereas the number of factors of 2 in $2n^3$ is one (the visible 2) *plus* a multiple of 3 (three times the number of 2s in the prime factorization of $n$).

Thus, irrationality is common, but we will soon see it is not a purely negative property. It is in fact useful and interesting. It allows us to create classes of irrational "integers" with properties similar to those of the ordinary integers. These so-called **algebraic integers** – a term we will later define in general – are useful because they allow us to split rational expressions, such as $x^2 - 2y^2$, into irrational factors that are more manageable. An example is worked out in the next section.

## Exercises

The Euclidean algorithm of Section 1.1 (repeated subtraction) also applies to a pair of numbers whose ratio is *irrational*, in which case the algorithm does not terminate. Indeed, Euclid stated nontermination of the algorithm as a *criterion for irrationality* in his *Elements*, Book X, Proposition 2. If we apply the Euclidean algorithm to the pair $(\sqrt{2}, 1)$ we find a new and enlightening proof that $\sqrt{2}$ is irrational. Here it will be convenient to work with ordered pairs $(a, b)$ in which $a > b$.

1. Check that the first two subtractions for the pair $\left(\sqrt{2}, 1\right)$ yield first the pair $\left(1, \sqrt{2} - 1\right)$ and then the pair $\left(2 - \sqrt{2}, \sqrt{2} - 1\right)$.
2. Show that the latter pair are in the *same ratio* as the first pair, and deduce that the Euclidean algorithm does not terminate.
3. Explain, on the other hand, why the algorithm terminates on any pair of numbers whose ratio is rational.

Nontermination of the Euclidean algorithm on the pair $\left(\sqrt{2}, 1\right)$ is clear because the process is *periodic*; the same ratio recurs every other step. This periodicity is reflected in the **infinite continued fraction** for $\sqrt{2}$.

4. Show that $\sqrt{2} + 1 = 2 + \left(\sqrt{2} - 1\right) = 2 + \frac{1}{\sqrt{2}+1}$.
5. By repeatedly replacing the denominator $\sqrt{2} + 1$ on the right-hand side of this equation by the whole right-hand side, show that

$$\sqrt{2} + 1 = 2 + \cfrac{1}{2 + \cfrac{1}{2 + \cfrac{1}{2 + \cfrac{1}{\ddots}}}}$$

and deduce that

$$\sqrt{2} = 1 + \cfrac{1}{2 + \cfrac{1}{2 + \cfrac{1}{2 + \cfrac{1}{\ddots}}}}$$

If one accepts that infinite continued fractions are meaningful (and they are!) then the latter expression is perhaps the clearest possible view of the irrationality of $\sqrt{2}$. Any irrational number is infinite in some sense, but a periodic infinity – the same thing over and over – is the easiest kind to grasp.

6. If one terminates the continued fraction $\sqrt{2}$ at its various levels, one obtains a sequence of fractions that converge to $\sqrt{2}$. Check that the first few fractions in this sequence are

$$\frac{2}{1}, \frac{3}{2}, \frac{7}{5}, \frac{17}{12}, \ldots$$

It happens that every other pair $x/y$ in this sequence satisfies $x^2 - 2y^2 = 1$.

## 1.5 The Equation $x^2 - 2y^2 = 1$

Trial and error shows that the smallest positive integer solution of $x^2 - 2y^2 = 1$ is $x = 3$, $y = 2$. We also notice that $1 = 3^2 - 2 \cdot 2^2$, viewed as a difference of two squares, has an irrational factorization

$$1 = (3 - 2\sqrt{2})(3 + 2\sqrt{2}).$$

Raising this equation to the $n$th power gives the equation

$$1 = 1^n = (3 - 2\sqrt{2})^n (3 + 2\sqrt{2})^n. \tag{*}$$

We now invoke a property of the numbers $a + b\sqrt{2}$, for integers $a$ and $b$, that holds only because $\sqrt{2}$ is irrational. Namely, the number $a + b\sqrt{2}$ "behaves the same" as its **conjugate** $a - b\sqrt{2}$ in the following sense:

**Multiplicative property of conjugation.** *The conjugate of a product is the product of the conjugates.*

*Proof.* Consider the product, for integers $a, b, c, d$,

$$(a + b\sqrt{2})(c + d\sqrt{2}) = ac + 2bd + (ad + bc)\sqrt{2}.$$

Its conjugate is $ac + 2bd - (ad + bc)\sqrt{2}$, which indeed is the product of the conjugates, $(a - b\sqrt{2})(c - d\sqrt{2})$.   □

It follows from this property that, if $x_n, y_n$ are the unique integers such that $(3 - 2\sqrt{2})^n = x_n - y_n\sqrt{2}$, then the product $(3 + 2\sqrt{2})^n$ of conjugates of factors on the left side equals the conjugate of the right side, $x_n + y_n\sqrt{2}$. We can now carry on from equation (*) as follows.

$$\begin{aligned} 1 &= (3 - 2\sqrt{2})^n (3 + 2\sqrt{2})^n \\ &= (x_n - y_n\sqrt{2})(x_n + y_n\sqrt{2}) \\ &= x_n^2 - 2y_n^2. \end{aligned}$$

Thus, *the integers $x_n, y_n$ defined by $(3 - 2\sqrt{2})^n = x_n - y_n\sqrt{2}$ are solutions of the equation $x^2 - 2y^2 = 1$ for all natural numbers $n$*. With a little more work, it can be shown that, apart from solutions obtained by changing the sign of $x_n$ or $y_n$, these are the only integer solutions.

The conjugation operation that unlocks these solutions is a property enjoyed by the numbers $a + b\sqrt{2}$ that give them "more structure" than ordinary integers. In some sense, the visible structure of the numbers $a + b\sqrt{2}$ reveals "hidden structure" in the ordinary integers. We will see more examples of this phenomenon in the next chapter. The numbers $a + b\sqrt{2}$ are among those we call algebraic integers, because they share many properties with the ordinary integers. We list the most fundamental shared properties in the next section.

## Exercises

The idea of encoding a solution $x = a$, $y = b$ of the equation $x^2 - 2y^2 = 1$ by the algebraic integer $a + b\sqrt{2}$ greatly clarifies the set of all solutions. It enables us to show that the solutions $x = x_n$, $y = y_n$ found above are in fact *all* the solutions (up to $\pm$ signs).

1. Show that if $a_1 + b_1\sqrt{2}$ and $a_2 + b_2\sqrt{2}$ encode solutions, then so does their product $(a_1 + b_1\sqrt{2})(a_2 + b_2\sqrt{2})$.

2. Show that if $a + b\sqrt{2}$ encodes a solution, so too does $(a + b\sqrt{2})^{-1}$.

It follows from these two exercises that we can speak of the *product* and *inverse* of ordered pairs $(a,b)$ such that $a^2 - 2b^2 = 1$; namely, by taking product and inverse of their "proxies" $a + b\sqrt{2}$. Next, we divide the solution pairs $(a,b)$ into classes: the *positive* pairs, for which $a > 0$, and the *negative* pairs, for which $a < 0$. We now concentrate on the positive pairs, since negative pairs are easily recovered from the positive pairs.

3. Show that if $(a,b)$ is a positive pair, then $a + b\sqrt{2}$ is positive, and so are the product and inverse of positive pairs.

By taking logarithms of the positive numbers $a + b\sqrt{2}$, we can now talk about *sum* and *negative* of positive solution pairs $(a,b)$. This view has the advantage of generating proxies of solution pairs by simple addition, and of ordering solution pairs by the size of the numbers $\log(a + b\sqrt{2})$.

4. Show that the numbers $\log(a + b\sqrt{2})$, for positive solution pairs $(a,b)$, include the numbers $\log(x_n + y_n\sqrt{2}) = n\log(3 + 2\sqrt{2})$ for each integer $n$.

The question that remains is whether the solution pairs $(a,b)$ include any other than those already found. This is where it helps to view $\log(a + b\sqrt{2})$ as the *size* of the solution $x = a$, $y = b$.

5. Suppose that $(a,b)$ is a positive solution pair *not* equal to any pair $(x_n, y_n)$, and suppose that $(x_m, y_m)$ is the solution pair "nearest" to $(a,b)$ in the sense that $|\log(x_m + y_m\sqrt{2}) - \log(a + b\sqrt{2})|$ is as small as possible. Deduce that there is a positive solution pair $(a',b')$ such that

$$0 < \log(a' + b'\sqrt{2}) < \log(3 + 2\sqrt{2}),$$

   which is a contradiction.

## 1.6 Rings

The system $\mathbb{Z} = \{\ldots, -2, -1, 0, 1, 2, 3, 4, \ldots\}$ of integers is the motivating example for the concept of **ring**, a set of objects with sum and product operations that share the following properties (also called the **ring axioms**) with $\mathbb{Z}$:

$$a + b = b + a \qquad\qquad ab = ba \qquad\qquad \text{(commutative laws)}$$
$$a + (b + c) = (a + b) + c \qquad a(bc) = (ab)c \qquad \text{(associative laws)}$$
$$a + 0 = a \qquad\qquad\qquad a \cdot 1 = a \qquad\qquad \text{(identity laws)}$$
$$a + (-a) = 0 \qquad\qquad\qquad\qquad\qquad \text{(inverse law)}$$
$$a(b + c) = ab + ac \qquad\qquad\qquad\qquad \text{(distributive law)}$$

Thus, in a ring $R$ we can add, subtract (by defining $a - b$ as $a + (-b)$), and multiply using the usual rules for calculation with whole numbers.[1] We can also define the concept of divisibility by saying that

$$b \text{ divides } a \Leftrightarrow a = bc \text{ for some } c \in R.$$

In fact, much of ring theory is motivated by trying to carry over Euclid's theory of primes to rings in general. This is not completely straightforward. For example, we might try to define a **prime** to be an element divisible only by itself and 1, but we know that even in $\mathbb{Z}$, a prime $p$, is divisible by $\pm 1$ and $\pm p$, due to the fact that $-1$ divides 1. Thus, the theory of divisibility is complicated by elements that divide 1, which are called **units**.

An example with *infinitely many* units is the ring

$$\mathbb{Z}[\sqrt{2}] = \{a + b\sqrt{2} : a, b \in \mathbb{Z}\}$$

that we used to find integer solutions of $x^2 - 2y^2 = 1$ in the previous section. The numbers $x_n - y_n \sqrt{2}$ defined there are units of $\mathbb{Z}[\sqrt{2}]$, because each of them divides 1; in fact

$$1 = (x_n - y_n \sqrt{2})(x_n + y_n \sqrt{2}).$$

Another complication is that unique prime factorization may fail, even with the usual proviso "up to unit factors." We will see an example in Chapter 2. An important example where it does *not* fail is the **polynomial ring** $\mathbb{Q}[x]$ of polynomials in $x$ with rational coefficients. It is clear that $\mathbb{Q}[x]$ is a ring

---

[1] The "usual rules" include many things that are not in the list of properties above. Indeed, the ring axioms have been chosen to be minimal, and some well-known properties can be derived from them only with some ingenuity. These properties include $a \cdot 0 = 0$ and $(-1) \cdot (-1) = 1$, which are posed as problems in Exercise 1 below. Also, it should be mentioned that our concept of ring is sometimes called a **commutative ring**. There is a more general concept of "ring" that allows noncommutative multiplication.

Figure 1.1 Simon Stevin (1548–1620). Licensed under Creative Commons Attribution-ShareAlike 4.0 International License.

under the usual sum and product of polynomials, and that any nonzero rational number is a unit. However, $\mathbb{Q}[x]$ has the division property in the following form:

**Division property for polynomials.** *If $a(x)$ and $b(x) \neq 0$ are in $\mathbb{Q}[x]$, then there are polynomials $q(x)$ and $r(x)$ in $\mathbb{Q}[x]$ such that*

$$a(x) = b(x)q(x) + r(x) \quad where \ \ \mathrm{degree}(r) < \mathrm{degree}(b).$$

*Proof.* This property falls out of the usual "long division" process for polynomials. If $\mathrm{degree}(a) < \mathrm{degree}(b)$, we can take $q(x) = 0$ and $r(x) = a(x)$, and we already have $\mathrm{degree}(r) < \mathrm{degree}(b)$.

If $a(x) = a_n x^n + a_{n-1} x^{n-1} + \cdots + a_0$ and $b(x) = b_m x^m + b_{m-1} x^{m-1} + \cdots + b_0$ and $n > m$, we begin by subtracting $\frac{a_n}{b_m} x^{n-m} b(x)$ from $a(x)$ to remove the term of highest degree in $a(x)$. Then we similarly remove the term of highest degree in $a(x) - \frac{a_n}{b_m} x^{n-m} b(x)$, and so on. This can continue as far as the degree $m$ term in $a(x)$, at which stage we have subtracted a polynomial $q(x)b(x)$ from $a(x)$ and the remainder $r(x)$ has degree less[2] than that of $b(x)$.                                                                    □

It was observed as early as 1585, by Simon Stevin, that this property gives a **Euclidean algorithm for polynomials**, enabling us to find their gcd.

It then follows, much as in Section 1.2, that in $\mathbb{Q}[x]$, the gcd of $a(x)$ and $b(x)$ is expressible in the form $a(x)m(x) + b(x)n(x)$ for some polynomials $m(x)$ and $n(x)$. This property in turn gives a prime divisor property and

[2] Notice that, to maintain that $\mathrm{degree}(r) < \mathrm{degree}(b)$ when $b(x)$ is a nonzero constant and $r(x) = 0$, we need the degree of 0 to be less than 0. The usual convention is that $\mathrm{degree}(0) = -\infty$.

unique prime factorization in $\mathbb{Q}[x]$ by the argument of Section 1.3. In this case, factorization is unique up to nonzero rational multiples of the factors.

**Notation and terminology.** For any ring $R$ we let $R[x]$ denote the **ring of polynomials** with coefficients in $R$ and "variable" or "indeterminate" $x$. We also let $R[a]$ denote the set of **values** obtained by substituting a value $a$ for $x$.

For example, $\mathbb{Z}[x]$ is the ring of polynomials with integer coefficients and $\mathbb{Z}\big[\sqrt{2}\big]$ is the set of values obtained by substituting $\sqrt{2}$ for $x$ in these polynomials. These values are precisely the numbers $a + b\sqrt{2}$ where $a, b \in \mathbb{Z}$.

## 1.6.1 Euclidean Domains

The Euclidean algorithm is not the only route to unique prime factorization, but it is common enough to be worth identifying the rings that possess such an algorithm. Guided by the two examples we have seen so far – $\mathbb{Z}$ and $\mathbb{Q}[x]$ – we capture such rings in the following definition, which says that a division property holds. It is also usual to assume that the ring contains no **zero divisors**; that is, nonzero elements whose product is zero.

**Definition.** A **Euclidean domain** $R$ is a ring with no zero divisors and with a **degree function** $d : R - \{0\} \to \mathbb{N}$ such that, for all $a, b \in R$ with $b \neq 0$, there are $q, r \in R$ such that

$$a = qb + r, \quad \text{with either } r = 0 \text{ or } d(r) < d(b).$$

In the case of the polynomial ring $\mathbb{Q}[x]$, the degree function is of course the polynomial degree. For $n \in \mathbb{Z} - \{0\}$, we can take $d(n) = |n|$.

In a Euclidean domain we can implement the Euclidean algorithm by repeated division with remainder. Given $a, b \in R$ with $b \neq 0$ we take $q, r$ so

$$a = qb + r, \quad \text{with } d(r) < d(b),$$

and let $a_1 = b$, $b_1 = r$. Then any common divisor of $a, b$ divides $r$ and hence also $a_1, b_1$, and we have $d(b_1) = d(r) < d(b)$. We repeat the process with $a_1, b_1$, similarly obtaining

$$a_1 = q_1 b_1 + b_2 \quad \text{with } d(b_2) < d(b_1) \text{ and } \gcd(a_1, b_1) = \gcd(a, b).$$

Then setting $a_2 = b_1$ gives $a_2, b_2$ with the same common divisors as $a_1, b_1$ – so $\gcd(a_2, b_2) = \gcd(a_1, b_1) = \gcd(a, b)$ – and so on. In general we get

$$a_i = q_i b_i + b_{i+1} \quad \text{with } d(b_{i+1}) < d(b_i) \text{ and } \gcd(a_i, b_i) = \gcd(a, b).$$

Since the $d(b_i)$ are decreasing natural numbers, the process ends necessarily when $b_n$ divides $a_n$ exactly, and hence $b_n = \gcd(a_n, b_n) = \gcd(a, b)$.

Thus, we have a Euclidean algorithm for gcd, and its consequences follow as they did for the Euclidean algorithm in $\mathbb{Z}$ or $\mathbb{Q}[x]$. In particular, we have the prime divisor property – if prime $p$ divides $ab$, then $p$ divides $a$ or $p$ divides $b$ – with the consequence that prime factorization is unique up to divisors of 1.

## Exercises

The ring axioms above are chosen to be economical, so they do not explicitly include some familiar properties. To see some properties that are implicit in the ring axioms, prove the following:

1. Uniqueness of additive inverse: If $a + b = 0$, then $b = -a$. Prove this by adding $-a$ to both sides and applying commutative and associative laws. Then prove in succession

$$a \cdot (-1) = -a, \quad a \cdot 0 = 0, \quad (-1) \cdot (-1) = 1.$$

Rings of polynomials will be particularly important in later chapters, so we take this opportunity to practice working with them.

2. Let $R[x_1, \ldots, x_n]$ denote the polynomials in $x_1, \ldots, x_n$ and coefficients in a ring $R$. Use induction on $n$ to prove that $R[x_1, \ldots, x_n]$ is a ring.
3. Use the Euclidean algorithm for polynomials in $\mathbb{Q}[x]$ to prove that

$$\gcd(x^3 - 2, x^2 - 1) = \gcd(x^2 - 1, x - 2) = \gcd(x - 2, 3) = 1.$$

4. Also, by observing the combinations of $x^3 - 2$ and $x^2 - 1$ computed by the algorithm, find polynomials $m(x)$ and $n(x)$ such that

$$(x^3 - 2)m(x) + (x^2 - 1)n(x) = 1.$$

(We revisit this example in the exercises to Section 4.4.)

The ring $\mathbb{Z}[\sqrt{2}] = \{r + s\sqrt{2}; r, s \in \mathbb{Z}\}$ is a Euclidean domain with degree function $d(r + s\sqrt{2}) = |r^2 - 2s^2|$. To prove this, first prove that $d$ extends to numbers $r + s\sqrt{2}$ for *rational* $r, s$ and is *multiplicative*.

5. Use the multiplicative property of conjugation from Section 1.5 to show

$$d\left((r + s\sqrt{2})(t + u\sqrt{2})\right) = d(r + s\sqrt{2})d(t + u\sqrt{2})$$

and observe that the proof holds for rational $r, s, t, u$.

Now suppose $a, b \in \mathbb{Z}[\sqrt{2}]$ with $b \neq 0$. We seek a "quotient" $q \in \mathbb{Z}[\sqrt{2}]$ with $d(a - qb) < d(b)$ or, by the multiplicative property of $d$, $d\left(\frac{a}{b} - q\right) < d(1) = 1$.

6. Given $a = r + s\sqrt{2}$ and $b = t + u\sqrt{2}$ in $\mathbb{Z}[\sqrt{2}]$, use the conjugate of $b$ to show that $\frac{a}{b} = v + w\sqrt{2}$ for rational $v, w$.

7. Choosing integers $x, y$ so that $|v - x| \leq \frac{1}{2}$ and $|w - y| \leq \frac{1}{2}$, show that $q = x + y\sqrt{2}$ satisfies $d\left(\frac{a}{b} - q\right) < 1$, as required.

It follows that unique prime factorization holds in $\mathbb{Z}[\sqrt{2}]$, up to divisors of 1. However, as we saw above, there are infinitely many divisors of 1 in $\mathbb{Z}[\sqrt{2}]$.

## 1.7 Fields

A **field** is a ring with the additional property that every nonzero element $a$ has a **multiplicative inverse**; that is, an element $a^{-1}$ such that $aa^{-1} = 1$. Thus, the **field axioms** are:

$$a + b = b + a \qquad\qquad ab = ba \qquad\qquad \text{(commutative laws)}$$
$$a + (b + c) = (a + b) + c \qquad a(bc) = (ab)c \qquad \text{(associative laws)}$$
$$a + 0 = a \qquad\qquad\qquad a \cdot 1 = a \qquad\qquad \text{(identity laws)}$$
$$a + (-a) = 0 \qquad\qquad aa^{-1} = 1 \text{ for } a \neq 0 \qquad \text{(inverse laws)}$$
$$a(b + c) = ab + ac \qquad\qquad\qquad\qquad \text{(distributive law)}$$

The multiplicative inverse[3] allows us to *divide* by any nonzero $a$; namely, by defining the **quotient** $b/a$ or $b \div a$ as $ba^{-1}$.

In a field we can add, subtract, multiply, and divide using the usual rules for calculation with numbers. The obvious example of a field is the system $\mathbb{Q}$ of rational numbers, but there are many other examples, both larger and smaller than $\mathbb{Q}$. Particularly important examples are the **finite fields** $\mathbb{F}_p$ for each prime number $p$. The elements of $\mathbb{F}_p$ are the **congruence classes** $[0], [1], [2], \ldots,$ $[p - 1]$, where

$$[a] = \{a + np : n \in \mathbb{Z}\}$$

and the sum and product operations are defined by

$$[a] + [b] = [a + b] \quad \text{and} \quad [a] \cdot [b] = [ab].$$

The set $[a]$ is called the **congruence class of** $a$ because its members are the integers $a'$ congruent to $a$ modulo $p$; that is, such that $p$ divides $a - a'$. Another way to put it is that $a$ and $a'$ leave the *same remainder* on division by $p$. We also write the congruence relation

---

[3] We can say *the* multiplicative inverse because it is easily proved to be unique. Namely, if $b$ is such that $ab = 1$, then we can prove that $b = a^{-1}$ by multiplying both sides of the equation $ab = 1$ by $a^{-1}$ and applying the commutative and associative laws.

$$a' \equiv a \pmod p.$$

The $+$ and $\cdot$ operations are *well-defined* on congruence classes; that is,

$$[a'] = [a] \text{ and } [b'] = [b] \quad \Rightarrow \quad [a' + b'] = [a + b] \text{ and } [a'b'] = [ab].$$

This is easily checked for sums. For products we suppose that $a' = a + mp$, $b' = b + np$ and calculate:

$$[a' \cdot b'] = [(a + mp)(b + np)] = [ab + (mb + na + mnp)p] = [ab].$$

These calculations show that $\mathbb{F}_p$ is a *ring* for any integer $p$. The ring properties are "inherited" from those of $\mathbb{Z}$. For example, $a + b = b + a$ in $\mathbb{Z}$, so

$$[a] + [b] = [a + b] = [b + a] = [b] + [a].$$

The crucial property that makes $\mathbb{F}_p$ a field – multiplicative inverse – is due to the primality of $p$. A multiplicative inverse of $[a]$ is a $[b]$ such that $[a][b] = [1]$. Since $[a][b] = [ab]$ and $[1] = \{1 + np : n \in \mathbb{Z}\}$, we seek a $b$ such that

$$ab = 1 + np \quad \text{for some } n \in \mathbb{Z}.$$

Equivalently, we seek $b$ such that $1 = ab - np$. This reminds us of something we saw in Section 1.3; $1 = \gcd(a, p) = ma + np$ for some $m, n \in \mathbb{Z}$. Thus we only have to rename the integer $n$ as $-n$, and we can take $b = m$.

A famous theorem of number theory can be reinterpreted in $\mathbb{F}_p$ (the original interpretation may be found in section 1.9):

**Fermat's little theorem.** *If $[a] \in \mathbb{F}_p$ is nonzero, then*

$$[a]^{p-1} = [1].$$

*Proof.* Consider the nonzero elements $[1], [2], \ldots, [p-1]$ of $\mathbb{F}_p$. For any nonzero $[a]$ the elements $[a][1], [a][2], \ldots, [a][p-1]$ are nonzero. And they are distinct, because we can recover $[1], [2], \ldots, [p-1]$ from $[a][1], [a][2], \ldots, [a][p-1]$ by multiplying the latter elements by the inverse $[b]$ of $[a]$.

Thus $[a][1], [a][2], \ldots, [a][p-1]$ are the same elements as $[1], [2], \ldots, [p-1]$, except possibly in a different order. At any rate, both sequences have the same product;

$$[1] \cdot [2] \cdots [p-1] = [a][1] \cdot [a][2] \cdots [a][p-1]$$
$$= [a]^{p-1}[1] \cdot [2] \cdots [p-1].$$

So if we cancel $[1], [2], \ldots, [p-1]$ from both sides by multiplying by their inverses, we get $[1] = [a]^{p-1}$, or $[a]^{p-1} = [1]$, as required.      $\square$

### 1.7.1 Finite Rings That Are Not Fields

The idea of congruence is not restricted to congruence mod $p$, where $p$ is prime. More generally, one has *congruence modulo n* (or mod $n$ for short) for any nonzero integer $n$, usually taken to be positive. The symbolism

$$a' \equiv a \pmod{n}$$

means that $n$ divides $a - a'$, and the numbers congruent to integer $a$, mod $n$, form the **congruence class of** $a$, mod $n$, also written $[a]$. By the same arguments as above, one finds that sum and product are well defined on congruence classes, and that the congruence classes form a ring under these operations.

However, if $n$ is not prime, the congruence classes mod $n$ do *not* form a field, because not every nonzero class $[a]$ has an inverse. A simple example consists of the congruence classes mod 6: $[0], [1], [2], [3], [4], [5]$. The classes $[2], [3]$ are not equal to $[0]$, yet

$$[2] \cdot [3] = [6] = [0].$$

Because of this, $[2]$ can have no inverse $[a]$, otherwise multiplying both sides by $[a]$ would give $[3] = [a][0] = [0]$, which is false. Similarly, $[3]$ can have no inverse.

In the general case where $n$ is not prime, say $n = lm$, one similarly finds that $[l], [m] \neq [0]$ but $[l][m] = [0]$, so neither $[l]$ nor $[m]$ can have an inverse among the congruence classes mod $n$. As mentioned in Section 1.6, nonzero elements whose product is zero, such as $[l]$ and $[m]$ here, are called **zero divisors**. We will see in Section 4.2 that zero divisors are the only things that prevent a ring being extended to a field, the way we extend the ring $\mathbb{Z}$ to the field $\mathbb{Q}$.

### Exercises

A ring may have no zero divisors without being a field: $\mathbb{Z}$, for example. But a *finite* ring with no zero divisors is a field, as can be seen by an argument rather like the proof of Fermat's little theorem. (We revisit this proof in Section 9.4.)

1. Suppose that $R$ is a ring with nonzero elements $a_1, a_2, \ldots, a_n$. If $a$ is any nonzero element of $R$, then $aa_1, aa_2, \ldots, aa_n$ are nonzero. Why?
2. Prove also that $aa_i = aa_j$ implies $a_i = a_j$.
3. Deduce that $aa_1, aa_2, \ldots, aa_n$ include the element 1, and hence that $a$ has a multiplicative inverse.

## 1.8 Factors of Polynomials

The proof of the division property of polynomials in $\mathbb{Q}[x]$, in Section 1.6, applies to polynomials with coefficients in any field $\mathbb{F}$, since it uses only the ordinary rules for addition, subtraction, multiplication, and division. With this remark we can extend a classical theorem about roots and factors, due to (Descartes, 1637, p. 159), from polynomials with numerical coefficients to polynomials with coefficients in any field $\mathbb{F}$. We denote this ring of polynomials by $\mathbb{F}[x]$.

**Factor theorem.** *If $f(x) \in \mathbb{F}[x]$ and $f(a) = 0$ for some $a \in \mathbb{F}$, then $f(x)$ has a factor $x - a$; that is, $f(x) = g(x)(x - a)$ for some $g(x) \in \mathbb{F}[x]$.*

*Proof.* According to the division property, division of $f(x)$ by $x - a$ gives

$$f(x) = g(x)(x - a) + r(x),$$

where $r(x)$ is a polynomial of degree less than that of $x - a$, so $r(x)$ is constant. Also, substituting $a$ for $x$ in this equation gives

$$f(a) = g(a) \cdot 0 + r(a) = r(a),$$

hence the constant polynomial $r(x) = 0$, since $f(a) = 0$.

Thus, $f(x) = g(x)(x - a)$.                                                                    □

**Corollary.** *A polynomial $f(x)$ of degree $n$ in $\mathbb{F}[x]$ has at most $n$ roots in $\mathbb{F}$.*

*Proof.* Each distinct root $a$ corresponds to a distinct factor $x - a$ in $f(x)$. Since $f(x)$ has degree $n$, $f(x)$ at most $n$ such factors, hence at most $n$ roots in $\mathbb{F}$.    □

This corollary, which is often attributed to Lagrange, simplifies the proof of another theorem of Fermat that we will see in Section 2.8.

### 1.8.1 Factor Theorem for Polynomials over a Ring

The factor theorem above is a neat application of division with remainder, but a stronger factor theorem can be obtained with only *multiplication*.

**Factor theorem over a ring.** *If $R$ is a ring, $f(x) \in R[x]$, and $f(a) = 0$ for some $a \in R$, then $f(x) = g(x)(x - a)$ for some $g(x) \in R[x]$.*

*Proof.* First observe that

$$(x - a)\left(x^{m-1} + x^{m-2}a + \cdots + xa^{m-2} + a^{m-1}\right)$$
$$= \left(x^m + x^{m-1}a + \cdots + xa^{m-1}\right) - \left(x^{m-1}a + \cdots + xa^{m-1} + a^m\right)$$
$$= x^m - a^m.$$

This calculation uses only the distributive property and basic properties of addition, so it holds in any ring.

Now suppose $f(x) = a_n x^n + \cdots + a_1 x + a_0$, where $a_0, \ldots, a_n \in R$. Then

$$f(x) - f(a) = a_n(x^n - a^n) + \cdots + a_1(x - a),$$

where $x - a$ is a factor of each term on the right-hand side, by the calculation above. Therefore,

$$f(x) - f(a) = g(x)(x - a), \quad \text{where} \quad g(x) \in R[x],$$

because all the coefficients of $g(x)$ arise from $a, a_0, \ldots, a_n$ by addition, subtraction, and multiplication, and hence they belong to $R$.

Finally, if $f(a) = 0$, we are left with $f(x) = g(x)(x - a)$.                    □

**Corollary.** *If* $f(x_1, \ldots, x_n) \in R[x_1, \ldots, x_n]$ *and* $f = 0$ *when* $x_i = a$ *for some* $a \in R$, *then* $f(x_1, \ldots, x_n) = g(x_1, \ldots, x_n)(x_i - a)$, *for some* $g(x_1, \ldots, x_n) \in R[x_1, \ldots, x_n]$.

*Proof.* View the multivariate polynomial $f(x_1, \ldots, x_n)$ as a single-variable polynomial $f^*(x_i)$, with coefficients in the ring

$$R^* = R[x_1, \ldots, x_{i-1}, x_{i+1}, \ldots, x_n],$$

which obviously contains $R$. Thus, we have $f^*(x_i) \in R^*[x_i]$ with $f^*(a) = 0$ for some $a \in R^*$.

It follows from the theorem above that

$$f(x_1, \ldots, x_n) = f^*(x_i) = g^*(x_i)(x_i - a), \quad \text{where} \quad g^*(x_i) \in R^*[x_i].$$

And, since $R^* = R[x_1, \ldots, x_{i-1}, x_{i+1}, \ldots, x_n]$, $g^*(x_i)$ is in fact a multivariate polynomial $g(x_1, \ldots, x_n) \in R[x_1, \ldots, x_n]$, as claimed.                    □

### Exercises

The high school algebra involved in the factor theorem over a ring has several interesting spinoffs.

1. Use the Corollary above to find three factors of

$$f(x, y, z) = xy^2 + yz^2 + zx^2 - x^2y - y^2z - z^2x,$$

and hence factorize this polynomial.
2. Find $g(x)$ such that $x^3 + a^3 = g(x)(x + a)$.
3. Show, by finding a variant of the factorization of $x^n - a^n$, that $x^m + a^m$ has a factor $x + a$ when $m$ is odd. What is $g(x)$ in this case?

These factorizations of polynomials influence the factorization of integers. In particular, they prevent certain numbers of the form $2^n + 1$ from being primes.

4. Explain why $2^m + 1$ is not prime when $m$ is odd.
5. Show in fact that $2^m + 1$ is prime only when $m$ is a power of 2.

The only known primes of the form $2^{2^h} + 1$ are those with $h = 0, 1, 2, 3, 4$. They are called **Fermat primes** because Fermat mistakenly conjectured that *all* numbers of the form $2^{2^h} + 1$ are prime. Euler found that this is not true for $h = 5$.

6. Check that 641 divides $2^{2^5} + 1$.

## 1.9 Discussion

For an excellent general history of algebra, from ancient times until the 20th century, see Katz and Parshall (2014). Their unifying theme is the solution of equations – "taming the unknown" – with generally no restriction on what type of solution is allowed. In this book we explore what happens when only integer solutions are sought, and more detailed historical information on this particular theme may be found in the book of Bashmakova and Smirnova (2000) and the chapter on algebra (also coauthored by Bashmakova) in Kolmogorov and Yushkevich (2001). Another useful history, which appeared after most of this book was written, is Gray (2018). Gray's book is particularly useful for its study of the theory of algebraic integers in the 19th century.

### 1.9.1 Arithmetic from Euclid to Fermat

The arithmetic thread of algebra, like many threads in mathematics, can be traced back to Euclid's *Elements*. The *Elements* is best known for its treatment of geometry, but its treatment of arithmetic is equally important. As we have seen, the *Elements* is the first known source of the Euclidean algorithm, the proof that there are infinitely many primes, the prime divisor property (and, implicitly, unique prime factorization), and the irrationality proof for $\sqrt{2}$.

Not so well known is Euclid's use of **induction**, in the form known as "infinite descent" or the **well-ordering** of $\mathbb{N}$. For example, in proving the existence of prime factorization, Euclid argues, as we did in Section 1.1, that natural numbers cannot decrease forever (*Elements*, Book VII, Proposition 31).

He implicitly does the same in assuming (Book VII, Proposition 1) that the Euclidean algorithm will terminate. For a long time mathematicians used induction for special tasks like this, often unconsciously, without realizing that induction is a foundation for virtually *everything* nontrivial in arithmetic. The first to realize this was Grassmann (1861), who gave inductive proofs of all the ring properties of $\mathbb{N}$. In the 1880s Peano and Dedekind used the idea to the hilt by making induction the fundamental axiom of arithmetic in the so-called **Peano axioms**.

The Euclidean algorithm also was not recognized as a fundamental part of arithmetic until the 19th century. This was partly because algorithms in general were not recognized as a special category of processes (indeed, this did not happen until the 20th century), but also because the Euclidean algorithm in particular could be subsumed under the topic of continued fractions. At least, this was the situation in Europe. The Euclidean algorithm had long been used in India, for purposes such as solving linear and quadratic Diophantine equations, under the name of the "pulveriser." For more on this story, see Weil (1984). In Europe, continued fractions were a topic that also encompassed quadratic Diophantine equations, especially the so-called **Pell equation** $x^2 - Ny^2 = 1$ for nonsquare positive integer values of $N$.

After discovering that $\sqrt{2}$ is irrational, Greek mathematicians studied the particular Pell equation $x^2 - 2y^2 = 1$, in order to approximate and better understand the quantity $\sqrt{2}$. They discovered the solutions $x = x_n$ and $y = y_n$ we found in Section 1.5 and called them "side and diagonal numbers" because the ratio $x_n/y_n$ approaches the ratio of the side to the diagonal of a square. (And perhaps also because they took $x_n, y_n$ as diagonal and side of an "approximate square," from which they constructed a better approximation $x_{n+1}, y_{n+1}$). Today we would relate their process to the continued fraction for $\sqrt{2}$.

The Indian mathematicians Brahmagupta (around 600 CE) and Bhaskara II (around 1150 CE) found methods for solving $x^2 - Ny^2 = 1$ for positive integers $N$, which, in the opinion of Weil (1984), came close to a general solution. Bhaskara II was able to solve the particularly hard case $N = 61$, whose smallest positive solution is $x = 1766319049$, $y = 226153980$. Fermat (1657) independently noted this case (great minds think alike?), so he probably had a general method for solving $x^2 - Ny^2 = 1$ as well. However, a general solution was first published by Lagrange (1768), using continued fractions. The key step is to show that the continued fraction for $\sqrt{N}$ is *ultimately periodic* for any positive integer $N$. This property can be seen in the continued fraction for $\sqrt{2}$ in exercise 5 of Section 1.4, where all quotients after the first are equal to 2.

Figure 1.2 Niccolò Fontana (1499–1557) and Gerolamo Cardano (1501–1576) (both public domain).

### 1.9.2 Algebra as Symbolic Calculation

The first step towards algebra as an independent discipline was the use of symbols to denote numbers, and extension of addition, multiplication, and so on from numbers to symbols. Rudimentary symbolism was used by Diophantus and later by al-Khwarizmi, whose work gave us the name "algebra" from the Arabic word "al-jabr" meaning something like "rearrangement of parts." (For a long time the word "algebra" was also used in Europe to mean resetting of broken bones.)

Algebra took off in Europe after Italian mathematicians in the early 16th century discovered the solution of cubic equations. Specifically,

$$x^3 = px + q \quad \text{has solution}$$

$$x = \sqrt[3]{\frac{q}{2} + \sqrt{\left(\frac{q}{2}\right)^2 - \left(\frac{p}{3}\right)^3}} + \sqrt[3]{\frac{q}{2} - \sqrt{\left(\frac{q}{2}\right)^2 - \left(\frac{p}{3}\right)^3}}.$$

Essentially, this solution was discovered by Scipione del Ferro soon after 1500. It was rediscovered by Niccolò Fontana (known as Tartaglia) in the 1530s, and published by Cardano (1545).

Until this time, algebraists interpreted algebra geometrically – for example, "completing the square" meant exactly that – and they used geometric proofs in the manner of Euclid. But the solution of cubic equations went beyond anything done by the Greeks, or anyone else, and it galvanized the European mathematicians. Cardano wrote:

Figure 1.3 Francois Viète (1540–1603) and René Descartes (1596–1650).
Bettman via Getty Images and via Getty Images, respectively.

In our own days Scipione del Ferro of Bologna has solved the case of the cube and
first power equal to a constant, a very elegant and admirable accomplishment.
Since this art surpasses all human subtlety and the perspicuity of mortal talent and
is a truly celestial gift and a very clear test of the capacity of men's minds, whoever
applies himself to it will believe that there is nothing he cannot understand.

By the end of the 16th century, European mathematicians were confidently
doing algebra by symbolic calculations alone, and indeed with much the same
notation used in high school algebra today. This notation was introduced by
Viète (1591), and Descartes (1637) used it to actually *reverse* the positions
of algebra and geometry, using algebra to prove results in geometry. This
**algebraic geometry** of Descartes paved the way for **calculus** and **differential
geometry**, in which the concept of "calculation" is extended to certain infinite
processes, such as infinite sums. This extension goes beyond what is normally
considered algebra, so we will not pursue it further in this book, but it can be
viewed historically as a natural outcome of the algebra of Viète and Descartes.

Indeed, we will not even pursue the operations of square root and cube
root that were of such interest to the early algebraists. Although we have used
numbers such as $\sqrt{2}$, and will continue to do so, we will not need to invoke any
infinite process – such as the continued fraction – in order to use $\sqrt{2}$ effectively
in algebra. We will see that to "know" $\sqrt{2}$ algebraically, it is enough to know
the concepts of sum, product, and congruence for polynomials. And these
concepts are virtually the same as the corresponding concepts for integers,
which we studied in Section 1.7.

### 1.9.3 The Binomial Theorem

One of the first important theorems of algebra is the **binomial theorem**:

$$(a+b)^n = a^n + na^{n-1}b + \frac{n(n-1)}{2}a^{n-2}b^2 + \cdots + nab^{n-1} + b^n,$$

where the coefficient of $a^{n-k}b^n$ is the so-called **binomial coefficient**

$$\binom{n}{k} = \frac{n(n-1)\cdots(n-k+1)}{k(k-1)\cdots 2\cdot 1} = \frac{n!}{(n-k)!\,k!},$$

also known as "$n$ choose "$k$."

The simplest examples are easily calculated:

$$(a+b)^1 = a+b,$$
$$(a+b)^2 = a^2 + 2ab + b^2,$$
$$(a+b)^3 = a^3 + 3a^2b + 3ab^2 + b^3$$
$$(a+b)^4 = a^4 + 4a^3b + 6a^2b^2 + 4ab^3 + b^4.$$

And the coefficients of any $(a+b)^n$ form the $n$th row of the so-called **Pascal's triangle**,

$$
\begin{array}{ccccccccccc}
 &  &  &  &  & 1 &  &  &  &  & \\
 &  &  &  & 1 &  & 1 &  &  &  & \\
 &  &  & 1 &  & 2 &  & 1 &  &  & \\
 &  & 1 &  & 3 &  & 3 &  & 1 &  & \\
 & 1 &  & 4 &  & 6 &  & 4 &  & 1 & \\
1 &  & 5 &  & 10 &  & 10 &  & 5 &  & 1
\end{array}
$$
$$\cdots\cdots\cdots\cdots\cdots\cdots\cdots,$$

each member of which, except the 1 at the end of each row, is the sum of the two above it.

"Pascal's triangle" was not discovered by Pascal. Examples of it may be seen in Chinese mathematics books centuries before Pascal. The Chinese used binomial coefficients to calculate numerical solutions of polynomial equations. Their process was later rediscovered under the name of "Horner's method" in Europe. However, Pascal (1654) was the first to systematically prove properties of the binomial coefficients, and to do so he used induction in the form commonly used today: with a "base step" and an "induction step."

For example, to prove the defining property of Pascal's triangle, one first observes that it is true for the second row 1  2  1 representing $(a+b)^2$; the 2 is indeed the sum of the two 1s above it. This is the base step.

Figure 1.4  Blaise Pascal (1623–1662). Licensed under Creative Commons Attribution 3.0 Unported License.

The induction step is to show that the coefficients in the $(n + 1)$st row are sums of those above them in the $n$th row. Well, the $(n + 1)$st row consists of the coefficients of $(a + b)^{n+1}$, and

$$(a + b)^{n+1} = a(a + b)^n + b(a + b)^n.$$

From the latter equation, we see that

$$\binom{n + 1}{k} = \text{coefficient of } a^k b^{n+1-k} \text{ in } (a + b)^{n+1}$$

$$= \text{coefficient of } a^k b^{n+1-k} \text{ in } a(a + b)^n$$

$$\qquad + \text{coefficient of } a^k b^{n+1-k} \text{ in } b(a + b)^n$$

$$= \text{coefficient of } a^{k-1} b^{n+1-k} \text{ in } (a + b)^n$$

$$\qquad + \text{coefficient of } a^k b^{n-k} \text{ in } (a + b)^n$$

$$= \binom{n}{k - 1} + \binom{n}{k},$$

which is the sum of the appropriate coefficients from the row above.

Fermat's "little theorem," proved in Section 1.7, was apparently discovered through a connection with properties of the binomial coefficients. Details may be found in Weil (1984), but the essential idea can be seen by looking at

$$2^p = (1 + 1)^p = 1^p + p \cdot 1^{p-1} + \frac{p(p - 1)}{2} \cdot 1^{p-2} + \cdots + p \cdot 1 + 1.$$

The binomial coefficients on the right-hand side are necessarily integers, so the denominator of each coefficient $\binom{p}{k}$ divides the numerator. It is clear from the formula for $\binom{n}{k}$ above that the numerator of $\binom{p}{k}$ contains the factor $p$ but the denominator does not. Therefore, by unique prime factorization, *p divides each term on the right-hand side, except the first and last.*[4] It follows that

$$2^p \equiv 2 \ (\text{mod } p).$$

Dividing each side by 2, which is valid when $p > 2$ because 2 has an inverse mod $p$, we find the special case of Fermat's little theorem:

$$2^{p-1} \equiv 1 \ (\text{mod } p).$$

Fermat in fact proved his little theorem in the form

$$a^p \equiv a \ (\text{mod } p),$$

which is valid for an arbitrary integer $a$.

### 1.9.4 Rings and Fields

Like any abstraction, the concept of field emerged only after several concrete instances of it had been observed: in this case, the field $\mathbb{Q}$ of rational numbers, and the **algebraic number fields** obtained from $\mathbb{Q}$ by throwing in roots of a polynomial in $\mathbb{Z}[x]$ and forming all possible sums, differences, products, and quotients. This was done by Galois in the late 1820s, as part of his theory of equations, which distinguishes the polynomial equations that have solutions "by radicals" (such as equations of degree 2 and 3) from those that do not (such as certain equations of degree 5). Remarkably, Galois also discovered the finite fields $\mathbb{F}_p$. At the time, fields were known as "domains of rationality" because they admit all the rational operations: addition, subtraction, multiplication, and division (by nonzero elements).

In the 1870s Dedekind singled out the algebraic number fields for special attention and gave them the name *Körper*. This is the German word for *body*, which Dedekind thought appropriate because closure under rational operations makes these domains in some sense complete and self-contained. Later, the word *Körper* became not *body* but *field* in English, though the ghost of *Körper* lingers in the letter $K$ often used to denote a field in English-language algebra books.

---

[4] This can be observed in rows 2, 3, and 5 of the small Pascal's triangle shown above. It is fun to continue the triangle as far as row 7, or row 11, to see how the phenomenon plays out for other prime-numbered rows.

Figure 1.5 Richard Dedekind (1831–1916) and Leopold Kronecker (1823–1891). Licensed under Creative Commons Attribution-ShareAlike 4.0 International License and courtesy of ETH-Zürich Bibliotek, respectively.

At about the same time Kronecker introduced the "finitary" approach to algebraic number fields, using congruence classes of polynomials in place of explicit irrational numbers. This approach, as we will see in Sections 4.3 and 4.4, follows the construction of finite fields via congruence of integers mod $p$. The only difference is that the Euclidean algorithm for integers is replaced by the Euclidean algorithm for polynomials. In Kronecker's construction, the prime number $p$ is replaced by a polynomial $p(x)$ that is "prime" (or **irreducible**) in the sense of having no factorization into polynomials of smaller degree.

The concept of ring emerged more slowly than the concept of field, from the concept of **algebraic integer**. Particular algebraic integers were used by Euler, who assumed that they behaved "like ordinary integers" to the extent of having unique prime factorization, as we will see in Section 2.9. But defining the general concept required some care: to obtain enough closure properties (under addition, subtraction, and multiplication) to ensure algebraic integers behave like ordinary integers, but not so many as to let them behave like rational numbers. Eventually, abstracting examples due to Dirichlet, Kummer, and Eisenstein, Dedekind came up with the appropriate definition in the 1870s. In the process, he shaped the future concept of ring.

The trick in defining algebraic integers is to take roots of polynomials in $\mathbb{Z}[x]$ (which define algebraic *numbers*), but to restrict the polynomials to those with leading coefficient 1: the so-called **monic** polynomials. Eisenstein (1850) had already proved that sum, difference, and product of such numbers is another number of the same type. So the collection of all algebraic integers is certainly a ring. However, Dedekind saw that a further restriction is needed to ensure anything like prime factorization: The algebraic integers should be confined to those of an algebraic number field.

Thus, instead of working with the ring of *all* algebraic integers, one normally chooses an algebraic number field $F$ and works with the ring of algebraic integers in $F$. We will see how this works out in Section 4.6. The term **ring** itself came later, as a shortening of the term *number ring* (in German, *Zahlring*) used by Hilbert (1897). The general theory of rings begins with Emmy Noether (1921), whose work will be discussed further in Chapters 5 and 9.

# 2

# Diophantine Arithmetic

## Preview

An equation is called **Diophantine** if it is a polynomial equation, with integer coefficients, for which integer solutions are sought. (Thus it is not really the equation that is "Diophantine," but the solutions.) Famous examples have already been mentioned in Sections 1.2 and 1.5. The name comes from Diophantus, who investigated such equations around 200 CE. He was actually more interested in *rational* solutions – which are usually easier to find – and it was Fermat who directed attention to integer solutions. Fermat was nevertheless greatly inspired by Diophantus, who made a few remarks on properties of integers, which Fermat realized were well worth pursuing.

They include properties of sums of two and four squares, and the special equation $y^3 = x^2 + 2$, whose solution $x = 5$, $y = 3$ was mentioned by Diophantus. Fermat saw that the key to understanding sums of two squares was finding the *primes* that are sums of two squares, which he claimed are those of the form $4n + 1$. He also made many other claims, among them that $x = 5$, $y = 3$ is the *only* positive solution of $y^3 = x^2 + 2$, and the famous *Fermat's last theorem*, stating that the equation $x^n + y^n = z^n$ has positive integer solutions only when $n \leq 2$.

These claims of Fermat inspired Euler, who managed to prove some of them in the next century. Among the tools Euler introduced were algebraic integers. These proved to be very fruitful for the future of number theory, and also for algebra.

## 2.1 Rational versus Integer Solutions

Perhaps the first, and certainly the most famous, Diophantine equation is

$$a^2 + b^2 = c^2,$$

relating the sides $a, b$, and hypotenuse $c$ of a right-angled triangle. Equally famous are some integer solutions of this equation, such as $(a, b, c) = (3, 4, 5)$ and $(a, b, c) = (5, 12, 13)$, called **Pythagorean triples**. Finding integer triples $(a, b, c)$ such that $a^2 + b^2 = c^2$ is essentially equivalent to finding *rational* solutions $x = a/c$ and $y = b/c$ of the equation:

$$x^2 + y^2 = 1.$$

Geometrically speaking, we seek rational points on the unit circle, and there is in fact a geometrical method of finding these points. (It was discovered by Diophantus, though he did not describe it geometrically.)

First we notice certain obvious rational points, such as $P = (-1, 0)$. If we connect $P$ to another rational point $Q$, then the line $PQ$ has rational slope. Not so obviously, the converse is true: if we draw a line through $P$ with rational slope $t$ (Figure 2.1), it meets the circle at another rational point $Q$.

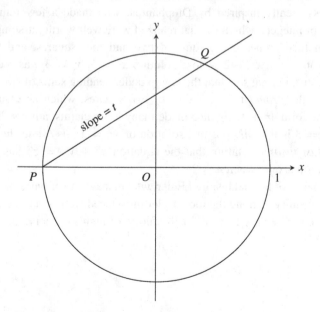

Figure 2.1 Finding the rational points on the unit circle.

There is an algebraic reason why $Q$ must be rational, but we need to find its specific coordinates in terms of $t$ in order to find a formula for Pythagorean triples. The reasoning and the calculation go as follows.

The equation of the line $PQ$ is $y = t(x+1)$, so we find the $x$-coordinates of its intersections with the circle by substituting $y = t(x+1)$ in $x^2 + y^2 = 1$. The result is the quadratic equation

$$(1+t^2)x^2 + 2t^2 x + (t^2 - 1) = 0$$

or, equivalently,

$$x^2 + \frac{2t^2}{1+t^2}x + \frac{t^2 - 1}{t^2 + 1} = 0.$$

Since $x = -1$ is a solution of this equation (corresponding to the point $P$), the left side has a factor $(x+1)$. The other factor is therefore $\left(x + \frac{t^2-1}{t^2+1}\right)$, which corresponds to the solution

$$x = \frac{1-t^2}{1+t^2}.$$

This is the $x$-coordinate of the point $Q$, and hence its $y$-coordinate is

$$y = t(x+1) = t\left(\frac{1-t^2}{1+t^2} + 1\right) = \frac{2t}{1+t^2}.$$

If we now substitute an arbitrary fraction $q/p$ for $t$, we find

$$x = \frac{a}{c} = \frac{p^2 - q^2}{p^2 + q^2} \quad \text{and} \quad y = \frac{b}{c} = \frac{2pq}{p^2 + q^2}.$$

This means that the original Pythagorean triple $(a, b, c)$ is of the form

$$a = (p^2 - q^2)r, \quad b = 2pqr, \quad c = (p^2 + q^2)r$$

for some natural numbers $p, q, r$,

a result originally found by Euclid.

A very similar argument enables us to find the rational points on the hyperbola $x^2 - 2y^2 = 1$. In fact, they are

$$x = \frac{1 + 2t^2}{1 - 2t^2}, \quad y = \frac{2t}{1 - 2t^2}.$$

But these are of little help in finding integer points. We can find by trial that $t = 1/2$ gives the smallest positive integer point, $(x, y) = (3, 2)$, and that $t = 2/3$ gives the second positive integer point $(x, y) = (17, 12)$. But there is no telling whether there are infinitely many integer points, or how to describe them. For this, as we saw in Section 1.5, algebraic integers are the key.

## Exercises

1. Show that the line of rational slope $t$ through the point $(-1,0)$ on the hyperbola $x^2 - 2y^2 = 1$ has a second intersection with the hyperbola at

$$x = \frac{1 + 2t^2}{1 - 2t^2}, \quad y = \frac{2t}{1 - 2t^2}.$$

2. Similarly find all rational points on the ellipse $x^2 + 3y^2 = 1$.

It should now be clear that this method will find all the rational points on a quadratic curve, as long as the curve has at least one rational point. But some quadratic curves have *no* rational points. An example is the circle $x^2 + y^2 = 3$.

3. Show that there are rational points on $x^2 + y^2 = 3$ only if there are integers $a, b, c$ such that $a^2 + b^2 = 3c^2$ and $\gcd(a, b, c) = 1$.
4. Show that any square is congruent to 1 or 0 mod 4.
5. Deduce that $a^2 + b^2 = 3c^2$ has no integer solution.

A direct approach to Euclid's solution of $a^2 + b^2 = c^2$ is possible with the help of unique prime factorization. The main idea is to seek *primitive* Pythagorean triples $(a, b, c)$, which are those for which the gcd of $a, b, c$ is 1.

6. By proving each square congruent to either 0 or 1, mod 4, show that exactly one of $a$ or $b$ is odd, and $c$ is odd, in a primitive Pythagorean triple $(a, b, c)$.
7. Assuming that $b$ is the even member of the triple, use the equation

$$\left(\frac{b}{2}\right)^2 = \frac{c - a}{2} \cdot \frac{c + a}{2}$$

and unique prime factorization to conclude that $\frac{c-a}{2} = p^2$ and $\frac{c+a}{2} = q^2$ for some integers $p, q$ with $\gcd(p, q) = 1$.
8. Solving for $a, b, c$, conclude that a primitive Pythagorean triple $(a, b, c)$ has the form $a = p^2 - q^2$, $b = 2pq$, $c = p^2 + q^2$, where $\gcd(p, q) = 1$.

## 2.2 Fermat's Last Theorem for Fourth Powers

Pythagorean triples are exceptional, according to the famous "last theorem" of Fermat. This theorem states that, for any integer $n > 2$, there are *no* positive integer triples $(a, b, c)$ with $a^n + b^n = c^n$. Although claimed by Fermat, the theorem was first proved by Wiles (1995). However, Fermat did give a proof for $n = 4$, which ingeniously uses Euclid's formula for Pythagorean triples and also Euclid's "infinite descent" form of induction. Here we give a shorter proof using the same general idea.

**Fermat's last theorem for fourth powers.** *If $a, b, c$ are positive integers, then*

$$a^4 + b^4 \neq c^4.$$

*Proof.* Suppose, for the sake of contradiction, that $a, b, c$ are positive integers with $a^4 + b^4 = c^4$. By dividing through by any common divisors, we can assume that the only common divisor of $a, b, c$ is 1, in which case the gcd of $a^2, b^2, c^2$ is also 1. Now the equation $a^4 + b^4 = c^4$ says that $(a^2, b^2, c^2)$ is a Pythagorean triple, so Euclid's formula gives

$$a^2 = p^2 - q^2, \quad b^2 = 2pq, \quad c^2 = p^2 + q^2, \quad \text{with} \quad \gcd(p, q) = 1.$$

This is because the gcd of $a^2, b^2, c^2$ is 1 and we can assume, without loss of generality, that $b^2$ is the even member of the triple.

We now view $a^2 = x$, $b = y$, and $c = z$ as integer solutions of the equation

$$x^2 + y^4 = z^4, \tag{*}$$

in which case $(x, y^2, z^2)$ is also a Pythagorean triple, with the gcd of $x, y^2, z^2$ equal to 1, and with $y^2$ even.

Then, by Euclid's formula again, we get positive integers $r, s$ with

$$x = r^2 - s^2, \quad y^2 = 2rs, \quad z^2 = r^2 + s^2, \quad \text{with} \quad \gcd(r, s) = 1.$$

The third equation shows that $(r, s, z)$ is yet another Pythagorean triple, whose members again have gcd 1, because a common prime divisor of $r, s$ would give a common prime divisor of $x, y^2, z^2$.

So, by Euclid's formula yet again, we get positive integers $u, v$, with gcd 1, such that either

$$r = u^2 - v^2 \text{ and } s = 2uv \quad \text{or} \quad r = 2uv \text{ and } s = u^2 - v^2.$$

In either case,

$$y^2 = 2rs = 4uv(u^2 - v^2) \quad \text{with} \quad \gcd(u, v) = 1.$$

Unique prime factorization then implies that $u$, $v$, and $u^2 - v^2$ are all squares, say,

$$u = l^2, \quad v = m^2, \quad u^2 - v^2 = l^4 - m^4 = n^2.$$

Now the last equation, $n^2 + m^4 = l^4$, *has the same form as equation* (*). Also, by tracing back, we see that $l \leq u < s < z$ or $l \leq u < r < z$. Thus, from any positive integer solution $z$ of (*) we find a *smaller* positive integer solution $l$. As Euclid might add, "which is impossible." Hence, there is no positive integer solution of $a^4 + b^4 = c^4$. □

## Exercises

The argument above proves incidentally that there is no positive integer solution of the equation $n^2 = l^4 - m^4$, and, hence, also no negative integer solution, since all the powers are even.

1. Deduce that there is no nonzero *rational* solution of $y^2 = 1 - x^4$, by showing that a nonzero rational solution would give a nonzero integer solution of $n^2 = l^4 - m^4$.

Jakob Bernoulli (1704) speculated that a similar argument might show there is no rational *function* solution of $y^2 = 1 - x^4$, which would explain why it seems impossible to "rationalize" $\sqrt{1 - x^4}$ by substituting a rational function for $x$, a problem he and other pioneers of calculus had met when trying to evaluate the integral $\int \frac{dx}{\sqrt{1-x^4}}$. Here is one such argument.

2. Show a solution of $y^2 = 1 - x^4$ by rational functions $y = y(t), x = x(t)$ gives a solution of $n^2 = l^4 - m^4$ by polynomials $l = l(t), m = m(t)$, $n = n(t)$.
3. Use unique prime factorization of polynomials, from Section 1.6, to rerun the argument in the previous exercise set with polynomials in place of integers. Hence, find a "Euclid's formula" for polynomials $a(t), b(t), c(t)$ satisfying $a(t)^2 + b(t)^2 = c(t)^2$.
4. Now carry over the argument for the impossibility of $n^2 = l^4 - m^4$ in nonzero integers to show its impossibility in nonzero polynomials.

## 2.3 Sums of Two Squares

In Book III, Problem 19, of his *Arithmetica*, Diophantus wrote:

> 65 is naturally divided into two squares in two ways, namely into $7^2 + 4^2$ and $8^2 + 1^2$, which is due to the fact that 65 is the product of 13 and 5, each of which numbers is the sum of two squares.

It is thought that Diophantus had in mind the two identities

$$(a^2 + b^2)(c^2 + d^2) = (ac - bd)^2 + (ad + bc)^2 \qquad (*)$$
$$= (ac + bd)^2 + (ad - bc)^2,$$

which give the two expressions for 65 as a sum of two squares by setting $a = 3$, $b = 2$, $c = 1$, $d = 2$. The general identities were later rediscovered, and laboriously proved, by Fibonacci (1225). Today, it is a routine exercise in high school algebra to confirm the identities, but they have a deeper meaning in connection with **complex numbers**.

If we interpret the "imaginary unit" $i = \sqrt{-1}$ as a unit vector perpendicular to the real number line, then each number of the form $a + bi$, where $a$ and $b$ are real, represents a point in the plane. The point $a + bi$ lies at distance $\sqrt{a^2 + b^2}$ from the origin, by the Pythagorean theorem, and we call this distance the **absolute value** $|a + bi|$ of $a + bi$. It is more convenient for number theory to use the square of the absolute value, $|a + bi|^2 = a^2 + b^2$, which we call the **norm**.

Thus, sums of two squares may be viewed as norms of complex numbers, and *the Diophantus identity* (*) *says that the norm is multiplicative.*

**Multiplicative property of the norm.** *For any complex numbers $a + ib$ and $c + id$*

$$|(a + bi)(c + di)|^2 = |a + bi|^2|c + di|^2.$$

*Proof.* We expand the product $(a + bi)(c + di)$, using $i^2 = -1$, and apply the definition of norm as the square of absolute value:

$$|(a + bi)(c + di)|^2$$
$$= |ac - bd + (ad + bc)i|^2$$
$$= (ac - bd)^2 + (ad + bc)^2 \quad \text{by definition of absolute value}$$
$$= (a^2 + b^2)(c^2 + d^2) \quad \text{by the Diophantus identity (*)}$$
$$= |a + bi|^2|c + di|^2 \quad \text{by definition of absolute value.} \qquad \square$$

The norm $|a + bi|^2$ is convenient for number theory because it is a natural number for natural numbers $a$ and $b$, though $|a + ib|$ has greater geometric

meaning. The absolute value acquires a multiplicative property from that of
the norm:

$$|(a + bi)(c + di)| = |a + bi||c + di|.$$

This says that multiplying the set of all complex numbers $a + bi$ by a
fixed complex number $c + di$ multiplies all distances by $|c + di|$, and hence
preserves shapes. (This idea is worked out more fully in the exercises below.)

Coming back to number theory, the complex numbers allow us to *factorize*
each sum of squares

$$a^2 + b^2 = (a - bi)(a + bi),$$

which draws our attention to numbers of the form $a + bi$ where $a$ and $b$ are
ordinary integers. These complex numbers, like the irrational real numbers
$a + b\sqrt{2}$ studied in Section 1.5, are "algebraic integers" in a sense we will
define in Section 4.6. They are called the **Gaussian integers**, and they form
a ring called $\mathbb{Z}[i]$. As in Section 1.5, we will find that these algebraic integers
have properties, such as conjugates and norm, that reveal hidden properties of
the ordinary integers. Not surprisingly, the Gaussian integers are particularly
good at revealing properties of sums of squares.

## Exercises

Although this is a book about algebra, sometimes algebra has a geometric
interpretation that is too useful to ignore. We have already seen an example
in the previous section. Another is the multiplicative property of the norm,
which has a geometric consequence about *preservation of shape* that will be
important in Section 2.5.

1. Observing that the distance between two points $v$ and $w$ in the plane of
   complex numbers is $|v - w|$, show that multiplication by $u$ sends $v$ and $w$
   to points whose distance apart is $|u||v - w|$.
2. Deduce that multiplication of all complex numbers by $u$ is a map of the
   plane that multiplies all distances by $|u|$, and hence *preserves shape*.
3. Show in particular that if $|u| = 1$, so $u = \cos\theta + i\sin\theta$, then
   multiplication by $u$ is a rotation about the origin $O$ through angle $\theta$.
4. Show, more generally, that multiplication by a general complex number
   $u = r(\cos\theta + i\sin\theta)$ is a combination of rotation about $O$ through angle
   $\theta$ and magnification by $r$.
5. Show that the unit circle $\{z \in \mathbb{C}: |z| = 1\}$ is divided into $n$ equal parts by
   the $n$ powers of $\zeta_n = \cos\frac{2\pi}{n} + i\sin\frac{2\pi}{n}$.
6. By solving the equation $z^3 - 1 = 0$, or otherwise, show that $\zeta_3 = \frac{-1+i\sqrt{3}}{2}$.

## 2.4 Gaussian Integers and Primes

The Gaussian integers $a - bi$ and $a + bi$ are called **conjugates** of each other, and we write $\overline{c + di}$ for the conjugate of $c + di$ for any ordinary integers $c$ and $d$. The conjugation operation has the following easily checked properties.

1. $\overline{(a + bi) + (c + di)} = \overline{(a + bi)} + \overline{(c + di)}$.
2. $\overline{(a + bi) \cdot (c + di)} = \overline{(a + bi)} \cdot \overline{(c + di)}$.
3. $(a + bi)\overline{(a + bi)} = a^2 + b^2 = |a + bi|^2$.

In particular, the second property is the multiplicative property of conjugation previously observed for the notion of conjugate used in Section 1.5. Notice also that properties 2 and 3 give another proof that the norm is multiplicative:

$$|(a + bi)(c + di)|^2 = (a + bi)(c + di)\overline{(a + bi)(c + di)} \quad \text{by property 3}$$
$$= (a + bi)(c + di)\overline{(a + bi)} \cdot \overline{(c + di)} \quad \text{by property 2}$$
$$= (a + bi)\overline{(a + bi)}(c + di)\overline{(c + di)}$$
$$= |a + bi|^2|c + di|^2 \quad \text{by property 3.}$$

There is in fact a general notion of conjugate for algebraic integers, and also a general notion of norm, for which analogues of all the above properties hold.

We can also use the norm to define the concept of **Gaussian prime**. We take the norm of a Gaussian integer to be its squared absolute value. Then a Gaussian integer is prime if it has norm greater than 1 and it is not the product of Gaussian integers of smaller norm. This generalizes an equivalent of the usual definition (in terms of divisors) for ordinary primes, and it is simpler to state because it avoids mentioning the eight trivial divisors of each Gaussian integer $\alpha$: namely, $\pm 1$, $\pm i$, $\pm \alpha$, and $\pm i\alpha$.

It follows, since the norm is an ordinary integer, that we have:

**Existence of prime factorization.** *Any Gaussian integer of norm greater than 1 has a Gaussian prime factorization.*

*Proof.* Suppose that $\alpha$ is a Gaussian integer of norm greater than 1. If $\alpha$ is a Gaussian prime, we are done. If not, suppose $\alpha = \beta\gamma$, where $\beta$ and $\gamma$ are Gaussian integers of smaller norm greater than 1. Repeat the argument with $\beta$. This produces a sequence of Gaussian divisors of $\alpha$ with decreasing norms. Since the norms are natural numbers, this process must terminate – necessarily with a Gaussian prime divisor $\rho$ of $\alpha$.

After finding a Gaussian prime divisor $\rho$ of $\alpha$ by this process, we repeat the process with the Gaussian integer $\alpha/\rho$. This leads to a Gaussian prime factorization of $\alpha$. $\qquad\square$

Proving *uniqueness* of this prime factorization is harder than proving existence, as it is with ordinary integers. However, the proof depends on the same basic idea: the *division property*. Before proving uniqueness, it may be worth looking at some Gaussian integers that factorize, and some that do not. The examples show how Gaussian factorization depends on factorization of the norm, thanks to the multiplicative property.

- $2 = (1 - i)(1 + i)$. The factors $1 - i$ and $1 + i$ are indeed Gaussian primes, because they each have a norm 2, which is not a product of smaller norms. Hence, neither $1 - i$ nor $1 + i$ is the product of smaller Gaussian integers.

  This example shows that an ordinary prime may not be a Gaussian prime. In fact, any prime that is the sum of two squares, $a^2 + b^2$, splits into Gaussian factors $a - ib$ and $a + ib$, and these factors are Gaussian primes because they have the (ordinary) prime norm $a^2 + b^2$.

- The ordinary prime 3 *is* a Gaussian prime because its norm $3^2$ splits into smaller factors only as $3 \cdot 3$, and 3 is not the norm of a Gaussian integer because $3 \neq a^2 + b^2$ for ordinary integers $a, b$.

- The Gaussian integer $3 + i$ has norm $10 = 2 \cdot 5$, so any Gaussian factors of $3 + i$ must must have norms $2 = 1^2 + 1^2$ and $5 = 2^2 + 1^2$. Those with norm 2 are the numbers $\pm 1 \pm i$ and those with norm 5 are either $\pm 2 \pm i$ or $\pm 1 \pm 2i$; all are Gaussian primes. By trial and error, we find $3 + i = (1 - i)(1 + 2i)$.

## Exercises

1. Use the existence of Gaussian prime factorization to prove that there are infinitely many Gaussian primes.

Finding the Gaussian prime factors of a Gaussian integer is not much harder than finding the ordinary prime factors of an ordinary integer. If the Gaussian integer is real, we first find its ordinary prime factors, then split those that are sums of two squares.

2. Find Gaussian prime factors of $30 = 2 \cdot 3 \cdot 5$.
3. Show in general that Gaussian prime factors of an ordinary integer are ordinary primes that are not sums of two squares and conjugate pairs of imaginary Gaussian primes.

If the Gaussian integer $\alpha$ has both real and imaginary parts, then its Gaussian prime factors correspond to ordinary prime factors of its norm, $|\alpha|^2$. If an ordinary prime factor of the norm is of the form $a^2 + b^2$, then the corresponding

Gaussian prime factor of $\alpha$ is one of $\pm a \pm bi$ or $\pm b \pm ai$, so we have to test a few (but not too many) possibilities.

4. Find a Gaussian prime factorization of $3 + 4i$, and hence show that it is the square of a Gaussian prime.
5. Find a Gaussian prime factorization of $7 - 11i$.

## 2.5 Unique Gaussian Prime Factorization

In $\mathbb{Z}[i]$ the division property is not as obvious as it is in $\mathbb{Z}$, but it is easy to prove by viewing $\mathbb{Z}[i]$ as a lattice of points in the plane. Figure 2.2 shows what they look like. They are shown as (mainly) white dots, with a few of black or gray. The black dots are multiples of $3 + i$ and the gray dot is the number $5 + 3i$.

Obviously, the points of $\mathbb{Z}[i]$ lie at the corners of a square lattice and, less obviously, so do the multiples of $3 + i$. This is because they result from multiplying the square lattice $\mathbb{Z}[i]$ by $3 + i$, which multiplies all distances by $|3+i|$ and hence preserves shape. Thus, the multiples of $3 + i$ lie at the corners of a lattice of squares of side length $|3+i|$. And the number $5 + 3i$, which lies in one of these squares, clearly lies at distance less than $|3+i|$ from the nearest corner.

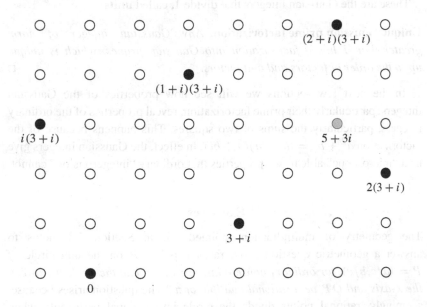

Figure 2.2 Multiples of $3 + i$ near $5 + 3i$.

More generally, the Gaussian integer multiples $\beta\mu$ of any Gaussian integer $\beta$ lie at the corners of a lattice of square of side length $|\beta|$, and any Gaussian integer $\alpha$ lies at a distance less than $|\beta|$ from the nearest multiple of $\beta$. This is due to the geometric fact (provable using the Pythagorean theorem) that the distance from a point in a square to the nearest corner is less than the side length. Translating this geometric fact into algebraic language, we have:

**Division property for Gaussian integers.** *For any Gaussian integers $\alpha$ and $\beta \neq 0$ there are Gaussian integers $\mu$ and $\rho$ such that*

$$\alpha = \beta\mu + \rho \quad and \quad |\rho| < |\beta|. \qquad \square$$

The division property now allows us to prove unique prime factorization for Gaussian integers, much as we did for ordinary integers. The only slight differences are:

- We must use division with remainder in the Euclidean algorithm rather than subtraction. But this still gives the essential consequence that the gcd of Gaussian integers $\alpha, \beta$ has the form $\mu\alpha + \nu\beta$ for some Gaussian integers $\mu, \nu$. This leads to the prime divisor property exactly as in Section 1.3.
- Prime factorization is unique only up to order and factors of $\pm 1$, $\pm i$, since Gaussian primes that divide each another may differ by the factors $\pm 1$, $\pm i$. These are the Gaussian integers that divide 1, called **units**.

**Unique Gaussian prime factorization.** *Any Gaussian integer of norm greater than 1 has a factorization into Gaussian primes, which is unique up to the order of factors and unit factors.* $\qquad \square$

In the next few sections we will see how properties of the Gaussian integers, particularly their prime factorization, reveal properties of the ordinary integers, particularly the sums of two squares. This happens because of the factorization $a^2 + b^2 = (a - bi)(a + bi)$. In effect, the Gaussian integers give us a "microscope" able to see properties that ordinary "integer vision" cannot.

## Exercises

The geometry of multiplication, pointed out in Section 2.3, helps to answer a geometric question about rational points $P$ on the unit circle: *if $P = (a/c, b/c)$, for ordinary nonzero integers $a, b, c$, can the angle $\theta$ between the $x$-axis and $OP$ be a rational multiple of $\pi$?* The question arises because, seemingly, rational points divide the circle into $n$ equal parts only when $n = 2$ or 4.

1. Suppose that $P = (a/c, b/c)$ is a point such that $OP$ makes an angle $2m\pi/n$ with the $x$-axis, where $n \neq 2, 4$. In other words,

$$\frac{a}{c} + \frac{b}{c}i = \cos\frac{2m\pi}{n} + i\sin\frac{2m\pi}{n}.$$

Deduce, from the geometric interpretation of multiplication, that

$$\left(\frac{a}{c} + \frac{b}{c}i\right)^n = 1, \quad \text{or} \quad (a + bi)^n = c^n.$$

2. Deduce from the equation $(a + bi)^n = c^n$, and unique Gaussian prime factorization, that the Gaussian prime factors of $(a + bi)^n$ differ from those of $c^n$ by at most the unit factors $\pm 1$ and $\pm i$.
3. Deduce in turn that the Gaussian prime factors of $a + bi$ differ from those of $c$ by at most unit factors.
4. Show, however, that $a + bi$ and $c$ differ by *more* than unit factors, so we have a contradiction.

The geometric view used to prove the division property in $\mathbb{Z}[i]$ can be adapted to prove the division property, and hence unique prime factorization, in $\mathbb{Z}[\sqrt{-2}]$, where $\mathbb{Z}[\sqrt{-2}]$ denotes the ring of numbers of the form $a + b\sqrt{-2}$ for $a, b \in \mathbb{Z}$.

5. For each nonzero $\beta \in \mathbb{Z}[\sqrt{-2}]$, the multiples $\beta\mu$ of $\beta \in \mathbb{Z}[\sqrt{-2}]$ by all $\mu \in \mathbb{Z}[\sqrt{-2}]$ form rectangles the same shape as those in $\mathbb{Z}[\sqrt{-2}]$. Why?
6. By considering the distance from any $\alpha \in \mathbb{Z}[\sqrt{-2}]$ to the nearest corner $\beta\mu$, prove the division property for $\mathbb{Z}[\sqrt{-2}]$.

## 2.6 Factorization of Sums of Two Squares

A sum of squares $a^2 + b^2$ of ordinary integers $a, b$ equals the Gaussian integer product $(a - bi)(a + bi)$, and we now know that $a - ib$ and $a + ib$ have unique Gaussian prime factorizations. We also know, by the multiplicative property of conjugates, that we can take the Gaussian prime factors of $a - bi$ to be the conjugates of the Gaussian prime factors of $a + bi$.

Then if we pair each Gaussian prime factor $a_k - b_k i$ of $a - bi$ with the Gaussian prime factor $a_k + b_k i$ of $a + bi$, we get the sum of squares $a_k^2 + b_k^2$, which is the norm of $a_k - b_k i$ (and of $a_k + b_k i$). Since $a_k - b_k i$ is a Gaussian prime, its norm is not a product of smaller norms: that is, $a_k^2 + b_k^2$ is not a product of smaller sums of squares.

Suppose that $a_k^2 + b_k^2$ is a product of ordinary primes $c_j$ that are *not* sums of squares. This is possible if one of $a_k, b_k$ equals 0, in which case

$$a_k^2 + b_k^2 = c^2$$

is the unique Gaussian prime factorization (up to unit factors), and hence $c$ is an ordinary prime that is a Gaussian prime. But if there are two or more $c_j$, then the product of the $c_j$ has a Gaussian prime factorization *different* from $(a_k - b_k i)(a_k + b_k i)$ (even up to unit factors), contrary to uniqueness.

Thus, the factors $a_k^2 + b_k^2$ of $a^2 + b^2$ are ordinary primes except when one of $a_k, b_k$ equals zero and the other equals an ordinary prime $c$. In this case, $c$ is a common divisor of $a - bi$ and $a + bi$, and hence of $a$ and $b$. So we are led to the theorem:

**Factorization of a sum of two squares.** *If $a$ and $b$ are ordinary integers with $\gcd(a, b) = 1$ then $a^2 + b^2$ is a product of ordinary prime sums of squares $a_k^2 + b_k^2$.*                                                                          □

This theorem explains the example of Diophantus, where 65 is the sum of squares in two ways, but it factorizes uniquely into prime sums of squares: namely, $13 = 3^2 + 2^2$ and $5 = 2^2 + 1^2$. A corollary of the theorem is a result discovered by Euler (1747):

**Divisors of a sum of squares.** *If $a$ and $b$ are ordinary integers with $\gcd(a, b) = 1$, then any integer divisor of $a^2 + b^2$ greater than 1 is a sum of two squares.*

*Proof.* Any divisor greater than 1 is a product of prime divisors, which, we have just seen, are the sums of squares $a_k^2 + b_k^2$. Such products are also sums of two squares, by the Diophantus identity of Section 2.3.                        □

## Exercises

Assuming unique prime factorization in $\mathbb{Z}[\sqrt{-2}]$, which was proved in the previous exercise set, we can now make a similar investigation of ordinary integers of the form $a^2 + 2b^2$ and their prime factors.

1. Check that the numbers 34, 54, and 102 are each of the form $a^2 + 2b^2$, and that each of their prime factors are also of this form.
2. Show the product of two numbers of the form $a^2 + 2b^2$ is also of this form.
3. Explain the latter result in terms of norm and conjugates of numbers of the form $a + b\sqrt{-2}$.

4. Prove that if $\gcd(a, b) = 1$, then any integer divisor of $a^2 + 2b^2$ greater than 1 is of the form $c^2 + 2d^2$.

We can also factorize members of $\mathbb{Z}[\sqrt{-2}]$ into primes *of* $\mathbb{Z}[\sqrt{-2}]$, using the natural number factors of their norms. For example, $1 + 2\sqrt{-2}$ has norm $9 = 3 \cdot 3$, which leads to the factorization $-(-1 + \sqrt{-2})^2$.

5. Factorize the number $3 + 2\sqrt{-2}$ into primes of $\mathbb{Z}[\sqrt{-2}]$.
6. Show that $5 + \sqrt{-2}$ is the cube of a prime of $\mathbb{Z}[\sqrt{-2}]$.

## 2.7 Gaussian Primes

Which ordinary primes $p$ are also Gaussian primes? If $p$ is a sum of two squares, $a^2 + b^2$ with $a, b \neq 0$, then $p$ has the factorization $(a - bi)(a + bi)$ into Gaussian integers of smaller norm, so $p$ is *not* a Gaussian prime. (However, the factors $a - bi$ and $a + bi$ of $p$ *are* Gaussian primes, because they have the ordinary prime $p$ as norm.)

Conversely, if $p = (a + bi)(c + di)$ is a factorization into Gaussian integers of smaller norm, then $a, b, c, d \neq 0$. Then, conjugating both sides of

$$p = (a + bi)(c + di) \quad \text{gives} \quad p = (a - bi)(c - di),$$

since $p$ is self-conjugate. Multiplying these two equations gives

$$p^2 = (a^2 + b^2)(c^2 + d^2),$$

so we necessarily have $p = a^2 + b^2 = c^2 + d^2$ by unique prime factorization in the natural numbers. Thus $p$ is a sum of two squares, and we have:

**Real Gaussian primes.** *An ordinary prime $p$ is a Gaussian prime if and only if $p$ is not a sum of two squares.* □

From here it is not far to reach a classification of all Gaussian primes:

**Gaussian primes.** *The Gaussian primes are of two types:*

- *Real primes $p$ that are not the sum of two squares, and the multiples of such primes by units.*
- *The factors $a - bi$ and $a + bi$ of real primes $p = a^2 + b^2$ with $a, b \neq 0$, and the multiples of such factors by units.*

*Proof.* If $a + bi$ is a Gaussian prime with $a, b \neq 0$, then $a - bi$ is also a Gaussian prime, since any factors of $a - bi$ give factors of $a + bi$ by conjugation.

Also, $\gcd(a,b) = 1$, otherwise $\gcd(a,b)$ is a real divisor of $a + bi$, again contrary to $a + bi$ being a Gaussian prime.

Then it follows from the factorization of the sum of two squares in the previous section that the norm $a^2 + b^2$ of $a + bi$ is an ordinary prime. If not, $a^2 + b^2$ splits into smaller norms $a_k^2 + b_k^2$, in which case $a + bi$ is not a Gaussian prime – because it splits into factors $a_k + b_k i$ of smaller norm.

Thus any Gaussian prime is either

- an ordinary prime, and hence not a sum of two squares, or its multiple by a unit, or else
- one of the Gaussian factors $a - bi, a + bi$ of an ordinary prime of the form $a^2 + b^2$ with $a, b \neq 0$.                                                              □

These results lead us to ask: *Which ordinary primes are sums of two squares?* And we are led to believe that the answer has something to do with the Gaussian primes.

## Exercises

Thanks to unique prime factorization in $\mathbb{Z}\left[\sqrt{-2}\right]$, established in the exercises to Section 2.5, we can classify the primes of $\mathbb{Z}\left[\sqrt{-2}\right]$ in the same way we classify Gaussian primes.

1. If an ordinary prime $p$ has a factorization into integers of smaller norm in $\mathbb{Z}\left[\sqrt{-2}\right]$,

$$p = \left(a + b\sqrt{-2}\right)\left(c + d\sqrt{-2}\right),$$

   show that $p = a^2 + 2b^2 = c^2 + 2d^2$.
2. Deduce that an ordinary prime $p$ is a prime of $\mathbb{Z}\left[\sqrt{-2}\right]$ if and only if $p$ is not of the form $a^2 + 2b^2$.
3. Hence, show that the primes of $\mathbb{Z}\left[\sqrt{-2}\right]$, other than $\pm$ ordinary primes, are the factors $a \pm b\sqrt{-2}$ of ordinary primes of the form $a^2 + 2b^2$.

## 2.8 Primes that Are Sums of Two Squares

Apart from the exceptional prime 2, all ordinary primes are odd. Of these, the sums of two squares are of the form $4n + 1$. This is because any odd sum of two squares must be the sum of an even square and an odd square; that is, of the form

$$(2a)^2 + (2b + 1)^2 = 4a^2 + 4b^2 + 4b + 1 = 4(a^2 + b^2 + b) + 1,$$

and this number is of the form $4n + 1$. A famous theorem of Fermat says that the converse is true: *any prime of the form $4n + 1$ is the sum of two squares.*

There is a simple proof of Fermat's theorem using Gaussian integers, due to Dedekind, though it depends on the following slightly technical lemma.

**Lagrange's lemma.** *If $p = 4n + 1$ is prime, then $p$ divides a number of the form $m^2 + 1$.*

*Proof.* By Fermat's little theorem (Section 1.7) we know $x^{p-1} \equiv 1 \pmod{p}$ for any $x \not\equiv 0 \pmod{p}$. This means that

$$0 \equiv x^{p-1} - 1 \equiv x^{4n} - 1 \equiv (x^{2n} - 1)(x^{2n} + 1) \pmod{4n + 1}$$

We also know, by Lagrange's theorem on polynomials over a field (Section 1.8) that a polynomial of degree $k$ has at most $k$ roots. Thus the $4n$ nonzero elements $a$ in $\mathbb{F}_{4n+1}$ can be separated into two groups of $2n$: the $2n$ roots of $x^{2n} - 1 = 0$ and the $2n$ roots of $x^{2n} + 1 = 0$.

If $a$ is any root of the latter equation, we have, since $4n + 1 = p$,

$$0 \equiv a^{2n} + 1 \equiv (a^n)^2 + 1 \pmod{p}.$$

In other words, if we set $m = a^n$, then $p$ divides $m^2 + 1$. □

**Fermat's two square theorem.** *If $p = 4n + 1$ is prime, then $p = a^2 + b^2$ for some $a, b \in \mathbb{Z}$.*

*Proof.* By Lagrange's lemma, there is an $m \in \mathbb{Z}$ such that

$$p \text{ divides } m^2 + 1 = (m - i)(m + i).$$

However, $p$ divides neither $m - i$ nor $m + i$ because neither $\frac{m}{p} - \frac{i}{p}$ nor $\frac{m}{p} + \frac{i}{p}$ is a Gaussian integer. So, by the prime divisor property for Gaussian integers (Section 2.5), we conclude that $p$ is *not* a Gaussian prime.

But then $p$ is a sum of two squares, as we saw in the previous section. □

## Exercises

It is interesting to see what happens when the above arguments are adapted to the case of an ordinary prime $p = 3n + 1$. We begin with a counterpart of Lagrange's lemma, obtained by imitating the proof above.

1. If $p = 3n + 1$, prove that $p$ divides a number of the form $m^2 + m + 1$.
2. Show that $m^2 + m + 1 = (m - \zeta)(m - \overline{\zeta})$, where $\zeta = \frac{-1+\sqrt{-3}}{2}$. Also show that $\overline{\zeta} \in \mathbb{Z}[\zeta]$.

3. Conclude that $p$ is *not* a $\mathbb{Z}[\zeta]$ prime, and hence is of the form $a^2 - ab + b^2$ (or, equivalently, $a^2 + ab + b^2$) if $\mathbb{Z}[\zeta]$ has unique prime factorization.

But does $\mathbb{Z}[\zeta]$ have unique prime factorization? We will see in the exercises to the next section.

## 2.9 The Equation $y^3 = x^2 + 2$

Euler (1770) gave a brilliant argument to show that the only positive integer solution of $y^3 = x^2 + 2$ is $x = 5$, $y = 3$. (Recall that this was the solution mentioned by Diophantus and claimed to be the only positive integer solution by Fermat.) Euler's idea was to use numbers of the form $a + b\sqrt{-2}$, for $a$, $b \in \mathbb{Z}$, which he guessed would behave like ordinary integers. This idea is correct, though not justified by him, and in this section we will see how and why it works.

Numbers of the form $a + b\sqrt{-2}$ are useful in studying the equation $y^3 = x^2 + 2$ because they allow it to be written

$$y^3 = \left(x - \sqrt{-2}\right)\left(x + \sqrt{-2}\right).$$

Assuming now that $x$ and $y$ are ordinary integers, Euler guessed that the gcd of the factors $x - \sqrt{-2}$ and $x + \sqrt{-2}$ on the right is 1. Assuming also that these factors behave like ordinary integers, he concluded that they must be cubes, since their product is the cube $y^3$.

If so, let

$$
\begin{aligned}
x - \sqrt{-2} &= \left(a - b\sqrt{-2}\right)^3 \\
&= a^3 - 3a^2b\sqrt{-2} + 3ab^2(-2) - b^3(-2)\sqrt{-2} \\
&= a^3 - 6ab^2 - (3a^2b - 2b^3)\sqrt{-2}
\end{aligned}
$$

Equating coefficients of $\sqrt{-2}$, we get

$$1 = 3a^2b - 2b^3 = b(3a^2 - 2b^2).$$

Thus, $b$ is an ordinary integer that divides 1, so $b = \pm 1$ and hence $b^2 = 1$. This makes $3a^2 - 2b^2$ (the other divisor of 1) equal to $3a^2 - 2$, which can only divide 1 if $a = \pm 1$. Since we want a positive value of $x = a^3 - 6ab^2$, we find by trial that $a = -1$, in which case $x = 5$.

Then we also have $y^3 = x^2 + 2 = 27$, so $y = 3$.

### 2.9.1 Justification

To clarify the ideas in Euler's proof, we use the name $\mathbb{Z}[\sqrt{-2}]$ for the ring of numbers $a + b\sqrt{-2}$ with $a, b \in \mathbb{Z}$, as we have already done for some time in exercises.

In $\mathbb{Z}[\sqrt{-2}]$ we call $a - b\sqrt{-2}$ the **conjugate** of $a + b\sqrt{-2}$ because it has the properties of conjugate listed in Section 2.4. This conjugate is in fact the usual complex number conjugate. It follows that the **norm** in $\mathbb{Z}[\sqrt{-2}]$ is

$$|a + b\sqrt{-2}|^2 = (a + b\sqrt{-2})(a - b\sqrt{-2}) = a^2 + 2b^2.$$

This implies that the only numbers of norm 1 in $\mathbb{Z}[\sqrt{-2}]$ are $\pm 1$, and hence that $\pm 1$ are the only **units** of $\mathbb{Z}[\sqrt{-2}]$.

It follows from the multiplicative property of conjugation that the norm is multiplicative, and hence that any divisors of $a + b\sqrt{-2}$ have norms that divide $a^2 + 2b^2$.

Now we can deal with the first important claim in Euler's proof: that the gcd of $x + \sqrt{-2}$ and $x - \sqrt{-2}$ is 1. This is actually *not* true for all ordinary integers $x$: for example, when $x = 2$, the gcd is $\sqrt{-2}$. But it is true if $x$ is odd, so first we have to check that $x$ is odd in any solution of $y^3 = x^2 + 2$. Well, if $x$ is even, then $x^2 + 2 = y^3$ is even, so $y$ is even. Suppose $y = 2m$, in which case $y^3 = 8m^3$ is divisible by 8. But $x^2 + 2 = (2n)^2 + 2 = 4n^2 + 2$ is not divisible by 4, let alone 8, so we have a contradiction.

Having established that $x$ is odd, we see that the norm $x^2 + 2$ of $x + \sqrt{-2}$ is odd, so $x + \sqrt{-2}$ is not divisible by any number with even norm. But any common divisor of $x - \sqrt{-2}$ and $x + \sqrt{-2}$ divides their sum, 2, whose norm 4 is not divisible by any odd natural number except 1. Thus, 1 is indeed the gcd of $x - \sqrt{-2}$ and $x + \sqrt{-2}$ when $y^3 = x^2 + 2$.

The second important claim is that $x - \sqrt{-2}$ and $x + \sqrt{-2}$ are both cubes. Given that $x - \sqrt{-2}$ and $x + \sqrt{-2}$ have no prime factor in common, this follows by distributing the prime factors of $y^3$, each of which occurs three times, between $x - \sqrt{-2}$ and $x + \sqrt{-2}$. Because, *assuming unique prime factorization*, each prime factor in $x - \sqrt{-2}$ occurs three times, and so does each prime factor in $x + \sqrt{-2}$. This implies that $x - \sqrt{-2}$ and $x + \sqrt{-2}$ are both cubes. (We leave it to the reader to check that the units $\pm 1$ are harmless.)

Thus, the second claim depends on proving unique prime factorization for $\mathbb{Z}[\sqrt{-2}]$. Fortunately, this is almost the same as the proof for $\mathbb{Z}[i]$ in Section 2.5. Recall that it was sufficient to prove the division property, which, for $\mathbb{Z}[\sqrt{-2}]$, reads:

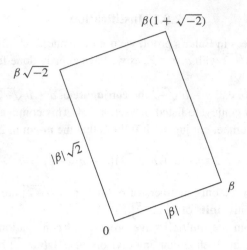

Figure 2.3 A typical rectangle in the lattice of multiples $\mu\beta$.

**Division property for** $\mathbb{Z}\left[\sqrt{-2}\right]$. *For any* $\alpha, \beta \in \mathbb{Z}\left[\sqrt{-2}\right]$ *with* $\beta \neq 0$ *there are* $\mu, \rho \in \mathbb{Z}\left[\sqrt{-2}\right]$ *such that*

$$\alpha = \beta\mu + \rho \quad \text{and} \quad |\rho| < |\beta|.$$

*Proof.* Recall also the argument for the division property: the set of multiples $\mu\beta$ form a lattice of the same shape as (in this case) $\mathbb{Z}\left[\sqrt{-2}\right]$, but magnified by $|\beta|$. This lattice consists of rectangles whose short side is $|\beta|$ and whose long side is $|\beta|\sqrt{2}$. Figure 2.3 shows one of them.

The argument concludes by saying that any point $\alpha$ in such a rectangle lies at distance $|\rho| < |\beta|$ from the nearest corner. This is not quite as clear as it was for the square lattice. But it can be seen by considering the point most distant from the corners – the center point – which is at distance

$$|\beta|\left[\left(\frac{1}{2}\right)^2 + \left(\frac{\sqrt{2}}{2}\right)^2\right] = |\beta|\left(\frac{1}{4} + \frac{2}{4}\right) < |\beta|,$$

by the Pythagorean theorem.                                                      □

## Exercises

The above argument does *not* work for $\mathbb{Z}\left[\sqrt{-3}\right]$.

1. Show that the division property fails in $\mathbb{Z}\left[\sqrt{-3}\right]$ by finding the distance from the center point of a rectangle with sides 1 and $\sqrt{3}$ to the nearest corner.

2. Show directly that unique prime factorization fails in $\mathbb{Z}\left[\sqrt{-3}\right]$ by finding two distinct prime factorizations of 4.

However, a geometric argument for the division property works in $\mathbb{Z}[\zeta]$, where

$$\zeta = \frac{-1 + \sqrt{-3}}{2},$$

and hence $\mathbb{Z}[\zeta]$ has unique prime factorization.

3. Show that the points 0, 1, and $1 + \zeta$ lie at the corners of an equilateral triangle in the plane of complex numbers, and hence that $\mathbb{Z}[\zeta]$ has the shape of a tiling of the plane by equilateral triangles.
4. Use an argument about shape to prove that the division property, and hence unique prime factorization, holds in $\mathbb{Z}[\zeta]$.
5. Show that the extra points of $\mathbb{Z}[\zeta]$ lie at the center points of the rectangles in $\mathbb{Z}\left[\sqrt{-3}\right]$.

We will see later (in Section 4.6) that it is appropriate to view the number $\zeta$ as an "algebraic integer."

## 2.10 Discussion

It is hard to exaggerate the influence of Diophantus on the later development of arithmetic and algebra. His work, the *Arithmetica*, contains only a series of examples of special problems and special solutions, yet the problems seem to have been chosen to illustrate general principles. His most important followers, Fermat and Euler, certainly understand them in this way. Fermat drew general conclusions from the special cases given by Diophantus and claimed to have proved them, though in most cases he did not reveal his proofs. Euler gave the first published proofs of several of Fermat's generalizations of Diophantus, and those he could not prove were a spur to later mathematicians. Indeed, even the results Euler proved were revisited by his successors – such as Lagrange, Gauss, Dirichlet, and Dedekind – in order to test new and more general methods.

The works of Diophantus may be seen in English translation in Heath (1964), and his influence on Fermat and Euler is traced in Weil (1984).

### 2.10.1 The Chord and Tangent Methods of Diophantus

Most problems in Diophantus are about finding rational solutions to equations, and it is the handful of problems about integer solutions that are most relevant to the subject of this book. Nevertheless, some of his methods for finding rational solutions are remarkable, and worth at least a brief discussion.

The first, called the **chord method**, was used in Section 2.1 to find the rational points on the circle, and hence to find a formula for Pythagorean triples. The outcome in this case was indeed a result about integer solutions, namely the solutions of $a^2 + b^2 = c^2$. However, this happened only because $a^2 + b^2 = c^2$ is *homogeneous* in the variables $a, b, c$, so an integer solution $(a, b, c)$ corresponds to a *rational* solution $x = a/c$, $y = b/c$ of $x^2 + y^2 = 1$.

The chord method works in general to find rational points on any quadratic curve $p(x, y) = 0$, where $p(x, y)$ is a quadratic polynomial with rational coefficients, provided the curve has a rational point $P$. We can then find any *other* rational point by drawing a line (a "chord") of rational slope $t$ through $P$ and finding the other point $Q$ where it meets the curve. Substituting the equation of the chord in $p(x, y) = 0$ gives a quadratic equation $q(x) = 0$, the two solutions of which represent $P$ and $Q$. These solutions also correspond to *factors* of $q(x)$; one of them is necessarily rational, since it represents $P$, so the other is also rational, which implies $Q$ is rational. Thus, we find, as in the case of the circle, that the rational points correspond one-to-one with the rational values of $t$.

The algebraic key to this argument is that $q(x) = 0$ is a quadratic equation with rational coefficients, and hence if one factor of $q(x)$ is rational, so is the other. But this works only for a polynomial $q(x)$ of degree 2. If we have a *cubic* curve $p(x, y) = 0$ and we know *two* rational points $P, Q$ on the curve, then a similar argument shows that a line through $P$ and $Q$ meets the curve in a third rational point $R$. (Because if $r(x)$ is a cubic polynomial with rational coefficients and two rational factors, then the third factor is also rational.)

But what if we know only *one* rational point $P$ on a cubic curve $p(x, y) = 0$? This is where geometry meets algebra in an interesting way, and it leads to the **tangent method**. Substituting the equation of a line of rational slope $t$ through $P$ in the equation $p(x, y) = 0$ gives a cubic equation $r(x) = 0$ with rational coefficients. So far, so good, but we know only *one* factor of $r(x)$: the rational factor corresponding to $P$. How can we turn one factor into two? The answer is by taking the line through $P$ to be the *tangent*.

The tangent at $P$ is the limit of a line through $P$ that meets the curve at a nearby point $P'$. While $P$ and $P'$ are separate, they correspond to separate, but nearly equal, factors of the the cubic polynomial $r(x)$. When $P'$ merges with $P$, bingo! The two factors become equal, so $r(x)$ now has two rational factors, and hence its third factor (corresponding to the other point $Q$ where the tangent at $P$ meets the curve) is also rational. This gives a second rational point, $Q$.

An example by Diophantus himself is in the *Arithmetica*, Book VI, Problem 18: the curve $y^2 = x^3 - 3x^2 + 3x + 1$. This equation has the obvious rational solution $x = 0$, $y = 1$. Diophantus makes the substitution $y = \frac{3}{2}x + 1$. Probably he did so because he could foresee the cancellation of $3x + 1$ that leads to the equation

$$0 = x^3 - \frac{21}{4}x^2 = x^2 \left( x - \frac{21}{4} \right),$$

and hence to another rational point with $x = 21/4$. But $y = \frac{3}{2}x + 1$ is none other than the tangent at the point $(0, 1)$, so the geometric argument above *explains* why it gives another another rational point (it also explains the two factors $x$). Whether or not Diophantus thought this way, the method is now called the "Diophantus tangent method."

### 2.10.2 Absolute Value

The Diophantus identity of Section 2.3,

$$(a^2 + b^2)(c^2 + d^2) = (ac - bd)^2 + (ad + bc)^2,$$

was of course interesting to him mainly as a fact about sums of squares. Its interpretation as a fact about complex numbers and their absolute values lay far in the future.[1] Nevertheless, Diophantus also had a two-dimensional interpretation in mind, since he associated any sum of squares $a^2 + b^2$ with the right-angled triangle with base $a$ and height $b$ (as would anyone whose mathematical education emphasized the Pythagorean theorem!).

The triangle interpretation adds a little content to the Diophantus identity, by associating with two given triangles a "product triangle" whose hypotenuse is the *product of* the hypotenuses of the given triangles. Figure 2.4 shows this interpretation visually. But this picture contains more than a product of lengths: there is also a *sum of angles*. The "product triangle" has base $ac - bd$ and height $ad + bc$, so the slope of its hypotenuse is

$$\frac{ad + bc}{ac - bd} = \frac{\frac{b}{a} + \frac{d}{c}}{1 - \frac{b}{a}\frac{d}{c}} = \frac{\tan\theta + \tan\varphi}{1 - \tan\theta \tan\varphi}, \tag{*}$$

where $\theta$ and $\varphi$ are the angles in the given triangles. If one recalls the formula from trigonometry,

$$\frac{\tan\theta + \tan\varphi}{1 - \tan\theta \tan\varphi} = \tan(\theta + \varphi),$$

[1] The interpretation of $a + ib$ as the point $(a, b)$ in the plane is credited to Wessel (1797), and it was rediscovered shortly thereafter by Argand and Gauss.

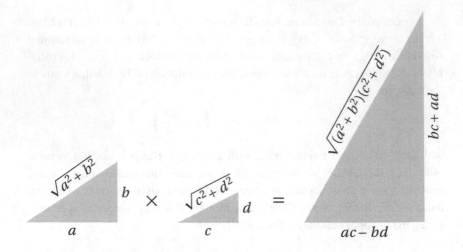

Figure 2.4  The Diophantus product of triangles.

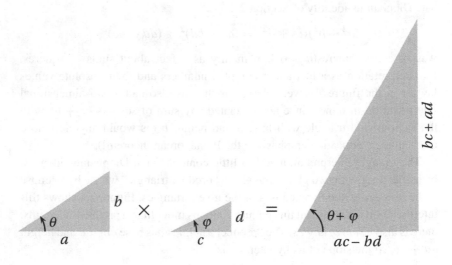

Figure 2.5  Angles in the product of triangles.

then (*) says that the angle of the "product triangle" is the sum of the angles in the given triangles. Figure 2.5 shows this additional information, which was apparently first observed by Viète (1615).

We can now see that the Diophantus identity has rich two-dimensional content. So, in a sense, the mathematical community was ready for the geometric interpretation of complex numbers by the time of Viète. The opportunity was missed, but we can only speculate why – perhaps the time was not yet ripe for an algebraic approach to geometry, which began only with Descartes (1637).

### 2.10.3  Norms

We have seen in this chapter that, for most applications to number theory, the squared absolute value, the **norm**, is more useful than the absolute value itself. This is because the norm is a natural number, so it reduces certain questions about algebraic integers to questions about natural numbers. In particular, we can exploit the fact that any set of natural numbers has a smallest member, and we can use prime factorization of natural numbers to discover factorizations of algebraic integers.

The parallel between natural number factorization and algebraic integer factorization occurs because of the **multiplicative property** of the norm:

$$\text{norm}(\mu \nu) = \text{norm}(\mu)\text{norm}(\nu).$$

In the case of Gaussian integers $\mu = a + ib$, $\nu = c + id$, this is just a restatement of the Diophantus identity. We saw in Section 2.4 how to use the multiplicative property to find factorizations of Gaussian integers and, in particular, to recognize Gaussian primes. It is a similar story in $\mathbb{Z}[\sqrt{-2}]$, where the norm is again just the square of the absolute value, and hence its multiplicative property follows from the multiplicative property of absolute value for complex numbers.

However, the multiplicative property of the norm extends to algebraic integers that are strictly real, such as the integers $a + b\sqrt{2}$ we investigated in connection with the special Pell equation $x^2 - 2y^2 = 1$. The norm of $a + b\sqrt{2}$ is defined to be the product of the **conjugates** $a + b\sqrt{2}$ and $a - b\sqrt{2}$, namely

$$\text{norm}(a + b\sqrt{2}) = (a + b\sqrt{2})(a - b\sqrt{2}) = a^2 - 2b^2.$$

Its multiplicative property then follows from the multiplicative property of conjugation, proved in Section 1.5.

The same is true of the algebraic integers $a + b\sqrt{N}$ for any nonsquare integer $N$. We define

$$\text{norm}(a + b\sqrt{N}) = (a + b\sqrt{N})(a - b\sqrt{N}) = a^2 - Nb^2,$$

and we can prove

$$\text{norm}\left((a + b\sqrt{N})(c + d\sqrt{N})\right) = \text{norm}(a + b\sqrt{N})\text{norm}(c + d\sqrt{N}).$$

This follows, again, by a multiplicative property of the conjugation operation, which in this case sends $a + b\sqrt{N}$ to $a - b\sqrt{N}$.

If we calculate the norms in the latter equation, we find, since

$$(a + b\sqrt{N})(c + d\sqrt{N}) = ac + bdN + (ad + bc)\sqrt{N},$$

an identity analogous to (and a generalization of) the Diophantus identity:

$$(ac + Nbd)^2 - N(ad + bc)^2 = (a^2 - Nb^2)(c^2 - Nd^2). \qquad (**)$$

The latter identity involves only real numbers, so it can be verified by expanding both sides and comparing the results. In fact it was discovered for positive nonsquare $N$ by Brahmagupta around 600 CE, and he used it to find solutions of the general Pell equation

$$x^2 - Ny^2 = 1.$$

In particular, it follows from (**) that from two solution pairs $(a,b)$ and $(c,d)$ of the Pell equation, we can calculate a third, namely $(ac + Nbd, ad + bc)$. For more on Brahmagupta and other Indian investigators of the Pell equation, see Weil (1984).

So far we have discussed only algebraic integers that satisfy quadratic equations, where the conjugates come in pairs $a + b\sqrt{N}, a - b\sqrt{N}$, and the norm is the product of the conjugates. This includes the cases of Gaussian integers and $\mathbb{Z}[\sqrt{-2}]$, where the product of conjugates equals the squared absolute value. Dedekind (1871) discovered that the norm concept can be generalized to more complicated algebraic numbers by generalizing the concept of conjugate. We will see in Chapter 6 that an algebraic number of degree $n$ has $n$ conjugates (including itself), and in Chapter 7 that its norm is the product of these conjugates. Again, the multiplicative property of the norm follows from a multiplicative property of conjugates.

However, there is another way to approach the multiplicative property. Readers familiar with the determinant concept from linear algebra will recall that the determinant function det also has a multiplicative property:

$$\det(AB) = \det(A)\det(B),$$

where $A$ and $B$ are square matrices of the same size and $AB$ denotes their matrix product. We will see in Section 6.4 that an algebraic integer $\alpha$ can be modeled by an integer matrix whose determinant (necessarily an ordinary integer) is the norm of $\alpha$. The multiplicative property of the norm then follows from the multiplicative property of det. We take the opportunity to offer a refresher course in determinants, including their multiplicative property, in Chapter 7.

# 3

# Quadratic Forms

## Preview

As we saw in the previous chapters, interesting Diophantine problems begin with quadratic equations. At the heart of many problems is the question of *prime* numbers given by quadratic formulas. Diophantus himself looked at primes of the form $x^2 + y^2$, and Fermat extended his investigation to primes of the form $x^2 + 2y^2$ and $x^2 + 3y^2$. In each of these cases, primes of the given "quadratic form" can also be expressed in a "linear form." An example is the odd primes of the form $x^2 + y^2$, which we know are those of the form $4n + 1$.

But Fermat was stymied by the form $x^2 + 5y^2$, whose primes seem to defy such a simple characterization. Lagrange probed this problem more deeply by introducing the notions of **equivalent** forms, and the **discriminant** of a form. He found that, when $k = 1, 2, 3$, all forms with the same discriminant as $x^2 + ky^2$ are equivalent, but there are inequivalent forms with the same discriminant as $x^2 + 5y^2$. This led Gauss (1801) to a thorough study of quadratic forms and their "composition," from which the concept of **Abelian group** emerged.

The general group concept will not be used explicitly in this book, but some Abelian groups will be important, and also their generalization to the concept of **module** studied in Chapter 8. Quadratic forms also meet the concept of module via the discriminant, which originally arose from linear maps of two variables but was later generalized to linear maps with any finite number of variables.

This chapter is somewhat less self-contained than the others in this book. It deals with concepts that are going to be superseded, so we sometimes give only sketchy explanations, with cross references to later parts of the book for the theory that replaces them.

## 3.1 Primes of the Form $x^2 + ky^2$

Fermat's theorem that odd primes of the form $x^2 + y^2$ are those of the form $4n + 1$ was the first of three theorems he discovered about primes of the form $x^2 + ky^2$, for $k = 1, 2, 3$. The first is stated in Fermat (1640), and the other two in Fermat (1654).

**Fermat's theorems on primes.** *If $p$ is any odd prime and $x$ and $y$ are integers, then*

*1. $p = x^2 + y^2 \Leftrightarrow p \equiv 1 \ (mod\ 4)$.*
*2. $p = x^2 + 2y^2 \Leftrightarrow p \equiv 1\ or\ 3 \ (mod\ 8)$.*
*3. $p = x^2 + 3y^3 \Leftrightarrow p \equiv 1 \ (mod\ 3)$.*

Fermat claimed to be able to prove these theorems, which was probably true, because Euler later found proofs of them using Fermat's methods. However, both Fermat and Euler got stuck on the form $x^2 + 5y^2$ (which is the next one of interest, since $x^2 + 4y^2$ is of the form $x^2 + y^2$). They left only the following conjectures, which show a puzzling distribution of primes in the congruence classes of 1, 3, 7, and 9 mod 20.

**Fermat's conjecture.** *If two primes are of the form $20n + 3$ or $20n + 7$, then their product is of the form $x^2 + 5y^2$.*

**Euler's conjecture.** *If $p$ is any prime, then*

$$p = x^2 + 5y^2 \Leftrightarrow p \equiv 1\ or\ 9 \pmod{20},$$

$$2p = x^2 + 5y^2 \Leftrightarrow p \equiv 3\ or\ 7 \pmod{20}.$$

The anomaly of $x^2 + 5y^2$ was an irritant to mathematicians until the late 19th century, and it was soothed only by the development of several new ideas. As (Weil, 1974, p. 104) said:

> When there is something that is really puzzling and cannot be understood, it usually deserves the closest attention because some time or other some big new theory will emerge from it.

### Exercises

1. Make a list of the odd primes less than 50; mark those of the form $4n + 1$, $8n + 1$, $8n + 3$ and $3n + 1$; and check that they can be written in the forms predicted by Fermat's theorems.
2. List the primes less than 100 that have the form $x^2 + 5y^2$, and use them to test the conjectures of Fermat and Euler.

Figure 3.1  Pierre de Fermat (1607–1665) and Leonhard Euler (1707–1783).
Public domain and via Getty Images, respectively.

## 3.2  Quadratic Integers and Quadratic Forms

As we saw in Section 2.9, Euler had great success using the "quadratic integers" $a + b\sqrt{-2}$ and, implicitly, exploiting their unique prime factorization. We also saw, in Section 2.8, that the Gaussian integers $\mathbb{Z}[i]$ are a very convenient tool for proving the first of Fermat's theorems. Their effectiveness is due to the factorization $x^2 + y^2 = (x - i)(x + i)$ and to unique prime factorization in $\mathbb{Z}[i]$.

We might similarly attack the second Fermat theorem using the factorization $x^2 + 2y^2 = (x - y\sqrt{-2})(x + y\sqrt{-2})$ and unique prime factorization in $\mathbb{Z}[\sqrt{-2}]$, which was established in Section 2.9. And a similar attack can be made on $x^2 + 3y^2 = (x - y\sqrt{-3})(x + y\sqrt{-3})$, admittedly working with a slightly larger ring of "integers" than $\mathbb{Z}[\sqrt{-3}]$ in order to secure unique prime factorization.

But this approach will definitely fail with $x^2 + 5y^2$ because the appropriate ring of "integers"

$$\mathbb{Z}[\sqrt{-5}] = \{a + b\sqrt{-5} : a, b \in \mathbb{Z}\}$$

does *not* have unique prime factorization. We can see this from the two factorizations of 6:

$$6 = 3 \cdot 2 = (1 - \sqrt{-5})(1 + \sqrt{-5}).$$

In $\mathbb{Z}[\sqrt{-5}]$, where the norm of $a + b\sqrt{-5}$ is $a^2 + 5b^2$, none of the numbers $2, 3, 1 - \sqrt{-5}$, or $1 + \sqrt{-5}$ splits into factors of smaller norm. This is because

Figure 3.2 Joseph-Louis Lagrange (1736–1813) and Carl Friedrich Gauss (1777–1855). Licensed under Creative Commons Attribution-ShareAlike 4.0 International License.

none of their respective norms – $2^2$, $3^2$, 6, 6 – splits into smaller norms in $\mathbb{Z}$, since neither 2 nor 3 is a norm. Thus $2, 3, 1 - \sqrt{-5}, 1 + \sqrt{-5}$ are "primes" of $\mathbb{Z}[\sqrt{-5}]$ in this sense, and yet the two "prime" factorizations of 6 are clearly different.

Nevertheless, this failure is revealing, as a possible explanation of the anomalous behavior of the form $x^2 + 5y^2$ discovered by Fermat. It is revealing to *us*, I should say, because in fact unique prime factorization was not identified as a key concept of arithmetic until Gauss (1801) explicitly stated it and gave a new proof in his *Disquisitiones Arithmeticae*. Moreover, by this time Gauss had also foreseen the possible *failure* of unique prime factorization, so he did not pursue the study of rings of "quadratic integers" such as $\mathbb{Z}[\sqrt{-5}]$.

Instead, Gauss made a deep study of **quadratic forms**: a field that had been opened up by Lagrange (1773).

## Exercises

The number $6 = 1^2 + 5 \cdot 1^2$ is of the form $a^2 + 5b^2$, with $a = 1, b = 1$. Other numbers of the form $a^2 + 5$ have similar behavior when factorized.

1. Show that $a^2 + 5$, for $a = 1, 2, 3, 4$, splits into ordinary primes *not* of the form $x^2 + 5y^2$, and that each of these numbers $a^2 + 5$ has nonunique "prime" factorization in $\mathbb{Z}[\sqrt{-5}]$.

Since unique prime factorization fails in $\mathbb{Z}[\sqrt{-5}]$ when "primes" are defined in terms of the norm, it must be that the division property fails when

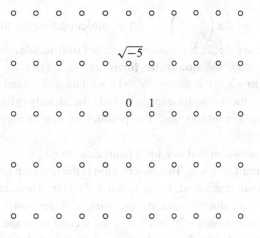

Figure 3.3 Geometric view of $\mathbb{Z}\left[\sqrt{-5}\right]$.

the size of remainders is measured by the norm (or, equivalently, by absolute value), since the division property implies unique prime factorization.

In fact we can also see that the division property fails by a direct geometric argument about the shape of rectangles in $\mathbb{Z}\left[\sqrt{-5}\right]$. Figure 3.3 shows some of these numbers as points in the plane $\mathbb{C}$. They clearly form rectangles of width 1 and height $\sqrt{5}$.

2. Explain why the multiples of any nonzero $\beta \in \mathbb{Z}\left[\sqrt{-5}\right]$ form rectangles of width $|\beta|$ and height $|\beta|\sqrt{5}$.
3. Show that a member of $\mathbb{Z}\left[\sqrt{-5}\right]$ inside one of these rectangles can lie at distance $> |\beta|$ from the nearest corner.
4. Deduce that the division property fails for $\mathbb{Z}\left[\sqrt{-5}\right]$.

## 3.3 Quadratic Forms and Equivalence

The reason for seeking primes of various forms, such as $x^2 + y^2$, goes back to Diophantus: it is the key to finding *all* numbers of the given form. And if one wants to know all numbers of the form $ax^2 + bxy + cy^2$, it may be useful to study other forms that represent the same set of numbers. This led Lagrange (1773) to consider mapping the variables $x, y$ to variables $x', y'$ given by equations of the form

$$x' = px + qy$$
$$y' = rx + sy, \quad \text{where} \quad p, q, r, s \in \mathbb{Z} \text{ and } ps - qr = \pm 1,$$

since $ps - qr = \det \begin{pmatrix} p & q \\ r & s \end{pmatrix} = \pm 1$ is the condition for the map to be a bijection of the set $\mathbb{Z} \times \mathbb{Z}$. Such maps are called **unimodular**.

Under this map, the **quadratic form** $ax^2 + bxy + cy^2$ becomes the **equivalent form** $a'x'^2 + b'x'y' + c'y'^2$. This relation is indeed an equivalence relation because the unimodular maps include the identity (giving reflexivity), inverses (giving symmetry), and the product of any two of them (giving transitivity).[1]

Lagrange discovered that any form equivalent to $ax^2 + bxy + cy^2$ has the same **discriminant** $b^2 - 4ac$. This observation is the first step towards deciding whether forms are equivalent, but it is not sufficient: There are inequivalent forms with the same discriminant. In particular, $x^2 + 5y^2$ and $2x^2 + 2xy + 3y^2$ both have the discriminant $-20$, but 7 is of the second form and not of the first.

On the bright side, the forms $x^2 + y^2$, $x^2 + 2y^2$, and $x^2 + 3y^2$ are each equivalent to all forms with the same discriminant. Lagrange proved this by a process giving each form with negative discriminant $D$ a **reduced form** $ax^2 + bxy + cy^2$ with $|b| \le a \le c$. From this it follows that

$$-D = 4ac - b^2 \ge 4a^2 - a^2 = 3a^2$$

and hence there are only finitely many reduced forms with given negative discriminant $D$. Certainly, $-D \ge 3a^2$ limits the values of $a$, hence also of $b$, and then also $-D = 4ac - b^2$ limits the values of $c$. So, one can find all reduced forms with a given negative discriminant $D$. This leads to the following results:

1. *All forms with discriminant $-4$ are equivalent to $x^2 + y^2$.*
2. *All forms with discriminant $-8$ are equivalent to $x^2 + 2y^2$.*
3. *All forms with discriminant $-12$ are equivalent to $x^2 + 3y^2$.*

But, as we have seen, not all forms with $D = -20$ are equivalent. *There are two inequivalent forms with discriminant $-20$, namely $x^2 + 5y^2$ and $2x^2 + 2xy + 3y^2$.*

Thus the equivalence theory of quadratic forms reveals a difference between the first three discriminants and the fourth: for the first three, the number of **equivalence classes** is 1, while for the fourth it is 2. The number of equivalence classes of forms with discriminant $D$ is called the **class number**, $h(D)$.

---

[1] Or, to put it in algebraic language, because the unimodular maps form a **group**. If the reader is not already familiar with the three defining properties of an equivalence relation, they may be found in Section 4.1.

It turns out that class number 1 implies unique prime factorization in the appropriate ring of quadratic integers.[2] But we can see immediately how the class number of $x^2 + y^2$ gives another proof of the Fermat two square theorem. This is basically Lagrange's proof, with a simplification of the crucial quadratic form due to Gauss (1801), article 182.

**Two square theorem revisited.** *If $p = 4n + 1$ is prime, then $p = x^2 + y^2$ for some $x, y \in \mathbb{Z}$.*

*Proof.* Thanks to Lagrange's lemma, used in the first proof of this theorem in Section 2.8, there is an integer $m$ such that $p$ divides $m^2 + 1$. Then $(m^2 + 1)/p$ is an integer, so $px^2 + 2mxy + \frac{m^2+1}{p}y^2$ is a quadratic form, with discriminant

$$b^2 - 4ac = (2m)^2 - 4p\frac{m^2 + 1}{p} = 4m^2 - 4(m^2 + 1) = -4.$$

Also, $px^2 + 2mxy + \frac{m^2+1}{p}y^2$ takes the value $p$ when $x = 1$ and $y = 0$.

Since this form has discriminant $-4$, it is equivalent to $x^2 + y^2$, so $x^2 + y^2$ also takes the value $p$. $\square$

## Exercises

Lagrange's result on forms of negative discriminant can also be used to prove the result in the exercises for Section 2.7 about primes $p$ of the form $3n + 1$.

1. Use Lagrange's result to show that the only reduced forms with $D = -3$ are $x^2 \pm xy + y^2$.
2. Deduce from exercise 1 of Section 2.8 that if $p = 3n + 1$, there is an $m \in \mathbb{Z}$ such that $px^2 + (2m + 1)xy + \frac{m^2+m+1}{p}y^2$ is a quadratic form, with $D = -3$.
3. Give another proof that the prime $p$ has the form $x^2 + xy + y^2$.

It is sometimes convenient to write each quadratic form as $ax^2 + 2bxy + cy^2$, and to take its discriminant to be $ac - b^2$, because then

$$(ax^2 + 2bxy + cy^2) = (x \ \ y) \begin{pmatrix} a & b \\ b & c \end{pmatrix} \begin{pmatrix} x \\ y \end{pmatrix}$$

---

[2] We will not explain this connection in detail, since we are going to drop the theory of quadratic forms in favor of the theory of quadratic integers. But, just briefly, class number 1 is equivalent to the statement that *all ideals are principal*, and we will show in Section 5.2, that the latter statement implies unique prime factorization.

and the discriminant is

$$ac - b^2 = \det\begin{pmatrix} a & b \\ b & c \end{pmatrix}.$$

Now suppose we have a change of variables $x' = px + qy$, $y' = rx + sy$, with matrix

$$T = \begin{pmatrix} p & q \\ r & s \end{pmatrix}.$$

4. Show that the form $a'x'^2 + 2b'x'y' + c'y'^2$ in the new variables has discriminant $\det(T)^2(ac - b^2)$, which equals $ac - b^2$ if the map is unimodular.[3]

## 3.4 Composition of Forms

Lagrange's theory of equivalent quadratic forms, and his discovery of the class number, revealed the hidden companion $2x^2 + 2xy + 3y^2$ of $x^2 + 5y^2$ and showed that these two forms represent the two equivalence classes of forms with discriminant $-20$. Lagrange also discovered that these two forms interact with each other by an operation later called **composition of forms**.

The first known example of the composition operation was the Diophantus identity of Section 2.3, which we now write as

$$(x_1^2 + y_1^2)(x_2^2 + y_2^2) = (x_1x_2 - y_1y_2)^2 + (x_1y_2 + y_1x_2)^2 = X^2 + Y^2, \quad (*)$$

which shows that any two expressions of the form $x^2 + y^2$ have a "composite" of the same form. Another example where two expressions of a certain form yield another of the same form was discovered by the Indian mathematician Brahmagupta around 600 CE:

$$(x_1^2 - ny_1^2)(x_2^2 - ny_2^2) = (x_1x_2 + ny_1y_2)^2 - n(x_1y_2 + y_1x_2)^2 = X^2 - nY^2, \quad (**)$$

Like $(*)$, $(**)$ can be proved by factorizing the left side and rearranging the factors, in this case as

$$\left[(x_1 - \sqrt{n}\,y_1)(x_2 - \sqrt{n}\,y_2)\right]\left[(x_1 + \sqrt{n}\,y_1)(x_2 + \sqrt{n}\,y_2)\right].$$

In each of these examples a form "composed with itself yields itself."

---

[3] This exercise assumes some knowledge of determinants and their *multiplicative property* $\det(AB) = \det(A)\det(B)$, at least for $2 \times 2$ matrices. However, in Chapter 7 we will develop their general theory from scratch, in order to put this knowledge on a sound foundation.

The identity (**) with $n = -5$ shows that the form $x^2 + 5y^2$ "composed with itself yields itself," but Lagrange found that the forms $x^2 + 5y^2$ and $2x^2 + 2xy + 3y^2$ have a more interesting interaction. First, the form $2x^2 + 2xy + 3y^2$ "composed with itself" does *not* yield itself, but rather $x^2 + 5y^2$, because

$$\left(2x_1^2 + 2x_1y_1 + 3y_1^2\right)\left(2x_2^2 + 2x_2y_2 + 3y_2^2\right) = X^2 + 5Y^2,$$

where $X = 2x_1x_2 + x_1y_2 + x_2y_1 - 2y_1y_2$ and $Y = x_1y_2 + x_2y_1 + y_1y_2$. And second, the form $x^2 + 5y^2$ composed with $2x^2 + 2xy + 3y^2$ yields $2x^2 + 2xy + 3y^2$ again, because

$$\left(x_1^2 + 5y_1^2\right)\left(2x_2^2 + 2x_2y_2 + 3y_2^2\right) = 2X^2 + 2XY + 3Y^2,$$

where $X = x_1x_2 - y_1x_2 - 3y_1y_2$ and $Y = x_1y_2 + 2y_1x_2 + y_1y_2$

These stunning feats of high school algebra can be reproduced almost mechanically by using the following factorizations in $\mathbb{Q}[\sqrt{-5}]$,

$$x^2 + 5y^2 = \left(x - y\sqrt{-5}\right)\left(x + y\sqrt{-5}\right)$$

$$2x^2 + 2xy + 3y^2 = 2\left[x + \frac{y}{2}\left(1 - \sqrt{-5}\right)\right]\left[x + \frac{y}{2}\left(1 + \sqrt{-5}\right)\right],$$

and rearranging so as to get conjugate factors, as above for (**).

More concisely, if the class of $x^2 + 5y^2$ is $A$ and the class of $2x^2 + 2xy + 3y^2$ is $B$, then we see that $A$ and $B$ have the following "multiplication table":

$$A^2 = A, \quad AB = BA = B, \quad B^2 = A.$$

Today, we would recognize this table as defining the two-element group with identity element $A$. It is now called the **class group** of $\mathbb{Q}[\sqrt{-5}]$ and is defined in an entirely different way. We can already see the advantage of working with the numbers $a + b\sqrt{-5}$, though admittedly the failure of unique prime factorization in $\mathbb{Z}[\sqrt{-5}]$ might be a problem.

It was perhaps this obstacle that Gauss (1801) was trying to avoid when he developed a comprehensive theory of composition of quadratic forms in his *Disquisitiones Arithmeticae*. He managed to define a composition operation for the inequivalent forms of any discriminant, but at the cost of enormous complications which virtually buried the underlying algebraic structure. In particular the **associativity** of the composition operation – $A(BC) = (AB)C$ – requires the derivation of 37 equations, most of which are left to the reader!

The easier theory of quadratic integers did not emerge until the 1870s, only after Dirichlet and Dedekind had used them to simplify the theory of quadratic forms as far as it would go. Eventually, quadratic forms receded when Dedekind (1871) found a theory of **algebraic integers** of any degree, overcoming the problem of prime factorization with his theory of **ideals**, and defining the class group in terms of ideals. We introduce ideals in Section 5.2.

## Exercises

The role of the form $2x^2 + 2xy + 3y^2$ can be observed in the numbers $a^2 + 5$ studied in exercise 1 of Section 3.2.

1. Check that the ordinary prime factors of the numbers $a^2 + 5$ for $a = 1, 2, 3, 4$ are all of the form $2x^2 + 2xy + 3y^2$.

Now let us see how Lagrange's identities arise from factorizations in $\mathbb{Z}[\sqrt{-5}]$.

2. Using the factorizations of $x^2 + 5y^2$ and $2x^2 + 2xy + 3y^2$ given above, show that

$$(x_1^2 + 5y_1^2)(2x_2^2 + 2x_2y_2 + 3y_2^2)$$
$$= 2(x_1 + y_1\sqrt{-5})\left[x_2 + \frac{y_2}{2}(1 + \sqrt{-5})\right]$$
$$\times (x_1 - y_1\sqrt{-5})\left[x_2 + \frac{y_2}{2}(1 - \sqrt{-5})\right].$$

3. By multiplying the first pair of factors and considering conjugates, show that the right-hand side equals

$$2\left[X + \frac{Y}{2}(1 + \sqrt{-5})\right]\left[X + \frac{Y}{2}(1 - \sqrt{-5})\right],$$

where $X = x_1x_2 - y_1x_2 - 3y_1y_2$ and $Y = x_1y_2 + 2y_1x_2 + y_1y_2$.

4. Finally, show that

$$2\left[X + \frac{Y}{2}(1 + \sqrt{-5})\right]\left[X + \frac{Y}{2}(1 - \sqrt{-5})\right] = 2X^2 + 2XY + 3Y^2.$$

## 3.5 Finite Abelian Groups

Before Dedekind's theory of algebraic integers was published, Kronecker (1870) published a very insightful paper proving what we now see as the **fundamental theorem on finite Abelian groups**. Motivated by the example of Gauss's composition of forms, Kronecker first defined such a group: a finite set $G$ of objects $g$ with an **identity** element 1 and an **inverse** element $g^{-1}$ for

Figure 3.4 A nonsquare rectangle.

each $g \in G$ with an operation (written as product) satisfying what we now call the **Abelian group axioms**:

$$g_1(g_2 g_3) = (g_1 g_2) g_3 \qquad \text{(associativity)}$$
$$g_1 g_2 = g_2 g_1 \qquad \text{(commutativity)}$$
$$g1 = g \qquad \text{(identity)}$$
$$gg^{-1} = 1. \qquad \text{(inverse)}$$

His theorem showed that a finite Abelian group is a product of the simplest Abelian groups, the **cyclic groups**. A group $C$ is finite and cyclic (and necessarily commutative) if it consists of the powers $1, c, c^2, \ldots, c^{n-1}$ of an element $c$ with $c^n = 1$. For example $c = -1$ gives a two-element cyclic group because $c^2 = 1$. The **direct product** $C \times D$ of groups $C$ and $D$ consists of the ordered pairs $(c, d)$ for $c \in C$ and $d \in D$ multiplied "coordinatewise"; that is, according to the rule $(c_1, d_1) \cdot (c_2, d_2) = (c_1 c_2, d_1 d_2)$. This is similarly true for the direct product of any number of groups. If a coordinate, say $c$, belongs to a cyclic group of $n$ elements, then one also has the rule $c^n = 1$.

Kronecker's theorem, in slightly modernized language, is the following.

**Structure of finite Abelian groups.** *Any finite Abelian group is either cyclic or the direct product of cyclic groups.*

Kronecker's proof is elementary but rather long, so I will be content to illustrate the theorem (which we do not need for this book) with a simple example: the **four-group**.

As its name suggests, the four-group has four elements; but it is *not* the cyclic group with four elements. The most graphic way to describe $G$ is as the group of symmetries of a nonsquare rectangle $ABCD$ (Figure 3.4). The symmetries of $ABCD$ are the moves that take the rectangle to a position that looks exactly the same, and they all result from flips over its horizontal and vertical axes of symmetry (dotted and dashed lines in the figure).

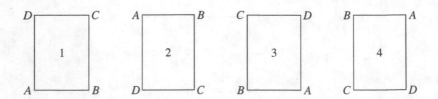

Figure 3.5 The four positions.

Informally, these are the moves by which a mattress may be turned and still fit on the bed. Experience shows there are four such moves, and they produce the positions of $ABCD$ shown in Figure 3.5.

1. No move.
2. Flip over the horizontal axis.
3. Flip over the vertical axis.
4. Two flips, one over each axis. (This move does not depend on the order of the flips and is a 180° rotation about the center point of the rectangle.)

Since the two flips commute, this group is Abelian. It is not cyclic because there is no single move which, when repeated, give all four positions. In fact, each move, done twice, returns the rectangle to its original position.

Thus, the four-group is an Abelian group of four elements that is not cyclic. It is the *direct product* of two-element cyclic groups $\{1, -1\}$, namely the "horizontal flip group" and the "vertical flip group." This is because each move in the four-group may be represented by an ordered pair $(a, b)$, where $a$ is 1 for no flip and $-1$ for a flip over the horizontal axis, and $b$ is 1 for no flip and $-1$ for a flip over the vertical axis.

## Exercises

Cyclic groups of one and two elements, namely $\{1\}$ and $\{1, -1\}$, occur in the real numbers $\mathbb{R}$, with ordinary multiplication as the group operation.

1. Show that a cyclic group of $n$ elements occurs in the complex numbers $\mathbb{C}$, with ordinary multiplication as the group operation.

Cyclic groups occur elsewhere in number theory; for example, in the finite fields $\mathbb{F}_p$ introduced in Section 1.7. It is a theorem of Gauss (1801), article 55, that the nonzero elements of $\mathbb{F}_p$ form a cyclic group under multiplication. An element $a \in \mathbb{F}_p$ whose powers are all the nonzero elements of $\mathbb{F}_p$ is called a **primitive root**.

2. Show that 2 is a primitive root for $\mathbb{F}_5$.
3. Find primitive roots for $\mathbb{F}_7$ and $\mathbb{F}_{11}$.

## 3.6 The Chinese Remainder Theorem

A natural example of a finite Abelian group that splits into a direct product occurs in the so-called Chinese remainder theorem, which is about multiplication mod $st$. Interestingly, this example is part of the splitting of a finite *ring* into a direct product. In Section 1.7 we discussed addition and multiplication mod $p$, where $p$ is a prime number. Here we extend the idea to addition and multiplication mod $k$, and then specialize to the case where $k = st$ where $s$ and $t$ are relatively prime; that is $\gcd(s, t) = 1$.

In the general case we have congruence classes mod $k$,

$$[a] = \{a + nk : n \in \mathbb{Z}\},$$

with the sum and the product of congruence classes given by $[a] + [b] = [a + b]$ and $[a] \cdot [b] = [ab]$. These operations are well-defined by the calculation in Section 1.7, where we also saw that the congruence classes inherit the ring properties from $\mathbb{Z}$. The ring of congruence classes mod $k$ is called $\mathbb{Z}/k\mathbb{Z}$, because it can be viewed as a **quotient** of the ring $\mathbb{Z}$ by the **ideal**[4] $k\mathbb{Z} = \{nk : n \in \mathbb{Z}\}$.

It is no longer always true that each nonzero class $[a]$ has an inverse. In fact, $[a]$ has an inverse if and only if $\gcd(a, k) = 1$, because

$$\gcd(a, k) = 1 \Leftrightarrow ma + nk = 1 \text{ for some } m, n \in \mathbb{Z}$$

$$\Leftrightarrow [a] \text{ has the inverse } [m], \text{ mod } k.$$

It follows that *the invertible elements* $[a]$ *mod $k$ form an Abelian group*, called $(\mathbb{Z}/k\mathbb{Z})^{\times}$, since the set of invertible elements clearly includes the identity class $[1]$, also the inverse of any member, and the product of any two members.

**Chinese remainder theorem.** *When $\gcd(s, t) = 1$, the ring $\mathbb{Z}/st\mathbb{Z}$ is isomorphic to $(\mathbb{Z}/s\mathbb{Z}) \times (\mathbb{Z}/t\mathbb{Z})$*

*Proof.* For any integer $a$ we will write the congruence classes of $a$ mod $st$, mod $s$, and mod $t$, respectively, as

---

[4] Quotients of rings by ideals will be introduced in Section 5.3. We mention them here only to explain where the notation $\mathbb{Z}/k\mathbb{Z}$ comes from.

$$[a]_{st} = \{a + nst : n \in \mathbb{Z}\},$$
$$[a]_s = \{a + ns : n \in \mathbb{Z}\},$$
$$[a]_t = \{a + nt : n \in \mathbb{Z}\}.$$

Consider the $st$ different remainders on division by $st$,

$$0, 1, 2, 3, \ldots, st - 1$$

and compare it with the first $st$ remainders on division by $s$, which consists of $t$ repetitions of the sequence

$$0, 1, 2, 3, \ldots, s - 1,$$

and also with the first $st$ remainders on division by $t$, which consists of $s$ repetitions of the sequence

$$0, 1, 2, 3, \ldots, t - 1.$$

Since $\gcd(p, q) = 1$, the latter two sequences coincide only at the first term, 0 (they next coincide with a 0 at position $st + 1$).

Thus, there is a bijection between each remainder on division by $st$ and the *pair* of remainders on division by $s$ and by $t$, respectively. In terms of congruence classes, this says that

$$\psi([a]_{st}) = ([a]_s, [a]_t) \quad \text{is a bijection.}$$

In fact, the bijection $\psi$ is a ring isomorphism, from the ring $\mathbb{Z}/st\mathbb{Z}$ of classes $[a]_{st}$ to the ring $(\mathbb{Z}/s\mathbb{Z}) \times (\mathbb{Z}/t\mathbb{Z})$ of ordered pairs $([a]_s, [a]_t)$, in which the sum and product are computed mod $s$ on the first coordinate and mod $t$ on the second coordinate.

Here is the calculation showing that $\psi$ preserves products:

$$\psi([aa']_{st}) = ([aa']_s, [aa']_t) \qquad\qquad\qquad \text{by definition of } \psi$$
$$= ([a]_s[a']_s, [a]_t[a']_t) \quad \text{by definition of mod } s \text{ and mod } t \text{ product}$$
$$= ([a]_s, [a]_t) \cdot ([a']_s, [a']_t) \qquad \text{by definition of product of pairs}$$
$$= \psi([a]_{st}) \cdot \psi([a']_{st}) \qquad\qquad\qquad \text{by definition of } \psi.$$

The calculation for sums is exactly the same, with $+$ in place of the $\cdot$ sign.

Now consider the subgroup $(\mathbb{Z}/st\mathbb{Z})^\times$ of the ring $\mathbb{Z}/st\mathbb{Z}$, consisting of the invertible elements $[a]_{st}$ under the product operation. The isomorphism $\psi$ sends these elements to the *invertible pairs* $([a]_s, [a]_t)$, and these are just the pairs where $[a]_s$ is in $(\mathbb{Z}/s\mathbb{Z})^\times$ and $[a]_t$ is in $(\mathbb{Z}/t\mathbb{Z})^\times$.

Thus, $(\mathbb{Z}/st\mathbb{Z})^\times$ is isomorphic to $(\mathbb{Z}/s\mathbb{Z})^\times \times (\mathbb{Z}/t\mathbb{Z})^\times$          □

**Exercises**

1. What is the group $(\mathbb{Z}/6\mathbb{Z})^\times$?
2. Show that $(\mathbb{Z}/2p\mathbb{Z})^\times$ is a cyclic group when $p$ is prime.

The number of invertible elements, mod $k$, is called $\varphi(k)$, and $\varphi$ is known as the **Euler phi function**. It is also the number of congruence classes $[a]_k$ for which $\gcd(a, k) = 1$, as we observed above. A fundamental property of the $\varphi$ function follows from the Chinese remainder theorem.

3. Deduce from the Chinese remainder theorem that if $\gcd(s, t) = 1$, then $\varphi(st) = \varphi(s)\varphi(t)$.
4. It follows from exercise 3 that $\varphi(m)$ may be calculated from a prime factorization of $m$. Thus, it suffices to know $\varphi(p^n)$ for each prime $p$. Show that $\varphi(p^n) = (p - 1)p^{n-1}$.

## 3.7 Additive Notation for Abelian Groups

In the last two sections we have given two historical examples of Abelian groups whose group operation is naturally viewed as "multiplication." We will later see many Abelian groups whose group operation is naturally viewed as addition. In fact, the oldest example of all – the integers – is another whose group operation is addition, not multiplication (because most integers have no multiplicative inverse, but they all have an additive inverse). Therefore, it is important to be aware of **additive notation** for Abelian groups, since this will often be the natural notation to use.

In additive notation, the group operation is written $+$, the identity element is written 0, and the inverse of element $g$ is written $-g$, so the defining properties or Abelian group axioms from Section 3.5 become:

$$g_1 + (g_2 + g_3) = (g_1 + g_2) + g_3 \qquad \text{(associativity)}$$
$$g_1 + g_2 = g_2 + g_1 \qquad \text{(commutativity)}$$
$$g + 0 = g \qquad \text{(identity)}$$
$$g + (-g) = 0. \qquad \text{(inverse)}$$

Since all of these properties hold for integer addition, it is now obvious why $\mathbb{Z}$ is an Abelian group under addition. Other examples are addition of congruence classes mod $n$, for any nonzero integer $n$. If $[g]$ denotes the congruence of integer $g$, mod $n$, then the set $\mathbb{Z}_n$ of these congruence classes inherits the Abelian group properties of addition from $\mathbb{Z}$. We saw this for the commutative property in Section 1.7, and it is similar for the others.

Alongside additive notation for individual groups, we have the notation for **direct sum** of two groups $G$ and $H$, $G \oplus H$, in place of the direct product notation $G \times H$. Since we write members of $G \times H$ as ordered pairs $(g,h)$, we can also write members of $G \oplus H$ as ordered pairs $(g,h)$, but "add componentwise,"

$$(g_1, h_1) + (g_2, h_2) = (g_1 + g_2, h_1 + h_2),$$

when $+$ is the group operation in $G$ and $H$. An alternative (common when working with vectors) is to write $(g,h)$ as $g\boldsymbol{i} + h\boldsymbol{j}$, then add the $\boldsymbol{i}$ and $\boldsymbol{j}$ components separately.

A variation of this idea that everyone knows is the notation $a + bi$ for complex numbers. Here $a$ and $b$ are real numbers, and the sum of two complex numbers is defined by

$$(a_1 + b_1 i) + (a_2 + b_2 i) = (a_1 + a_2) + (b_1 + b_2)i.$$

This shows that the Abelian group of complex numbers $\mathbb{C}$, under addition, is the same as the direct sum $\mathbb{R} \oplus \mathbb{R}$ of the real numbers with itself.

## Exercises

An Abelian group that should be viewed *both* multiplicatively and additively came up in the exercises to Section 1.5. There we studied solutions $x = a$, $y = b$ of the equation $x^2 - 2y^2 = 1$ through their "proxies" $a + b\sqrt{2}$ and (for $a > 0$) $\log\left(a + b\sqrt{2}\right)$.

1. Show that the numbers $a + b\sqrt{2}$, where $a$ and $b$ are integers and $a^2 - 2b^2 = 1$, form an Abelian group under multiplication.
2. Show that the subset of these numbers for which $a > 0$ is also an Abelian group.
3. Show that numbers $\log\left(a + b\sqrt{2}\right)$, where $a > 0$ and $b$ are integers such that $a^2 - 2b^2 = 1$, form an Abelian group under addition.

## 3.8 Discussion

The topic of quadratic forms sketched in this chapter was an important stepping stone from classical arithmetic to algebraic number theory and to some concepts of commutative algebra, such as Abelian groups and the discriminant.

However, we view it mainly as a topic of historical interest, which motivates later developments but is superseded by them. In particular, in the rest of the book the concept of quadratic form itself will be replaced by the more enlightening concept of quadratic integer.

### 3.8.1  Quadratic Forms and Quadratic Integers

Quadratic forms have played an important role in the development of algebra. On the one hand, they had been studied since ancient times in quadratic Diophantine equations, such as the Pythagorean equation and the Pell equation, so they are of abiding interest. On the other hand, they have problematic behavior: there are inequivalent forms with the same discriminant, which leads to failure of unique prime factorization in the corresponding quadratic integers. This created a demand for some *replacement* for unique prime factorization. Kummer expressed the demand, or hope, as follows:

> It is greatly to be lamented that this virtue of the real numbers [that is, ordinary integers] to be decomposable into prime factors, always the same ones for a given number, does not also belong to the complex numbers [that is, algebraic integers]; were this the case, the whole theory, which is still laboring under such difficulties, could easily be brought to its conclusion. For this reason, the complex numbers we have been considering seem imperfect, and one may well ask whether one could not look for another kind which would preserve the analogy with the real numbers with respect to such a fundamental property.
> *(Translation by Weil (1975) from Kummer (1844)).*

Dedekind felt the loss of unique prime factorization as keenly as Kummer, so he had the utmost admiration for Kummer's way out of the difficulty.

> But the more hopeless one feels about the prospects of later research on such numerical domains [of algebraic integers], the more one has to admire the steadfast efforts of Kummer, which were finally rewarded by a truly great and fruitful discovery.
> *(Translation in (Dedekind, 1996, p. 56)).*

Kummer's way out was the introduction of what he called "ideal numbers," but he did not define what these "numbers" were. Dedekind, inspired by Kummer, gave the "ideal numbers" a concrete existence as what he called **ideals**. Today, ideals are the first topic one meets in ring theory, though usually without learning where they came from.

Ideals had a long and difficult birth. As we have seen, mathematicians were reluctant to give up quadratic forms and accept quadratic integers. In fact, the

Figure 3.6  Ernst Eduard Kummer (1810–1893). Courtesy of the Mathematische Gesellschaft in Hamburg.

move to quadratic integers was achieved only with Dedekind's construction of a general theory of algebraic integers of arbitrary degree.

### 3.8.2  From Quadratic Forms to Algebraic Integers

The problematic behavior of quadratic forms was initially met with complicated and ingenious, but still traditional, algebra. First there was Lagrange's method of reduction, which had some success with forms of negative discriminant. Then there was Gauss's theory of composition of forms, in Gauss (1801), which could handle forms of arbitrary discriminant but was formidably complicated. Gauss's theory, which fills over 200 pages in his *Disquisitiones Arithmeticae*, was impenetrable to most mathematicians and the *Disquisitiones* gained the reputation as a "book with seven seals."

After decades of work, Dirichlet (1863) produced a simplified, and certainly more readable, version of Gauss's theory. But composition of forms was still a mystery. As we have seen, the Abelian group structure of composition was not brought to light until Kronecker (1870) made it explicit and proved the fundamental theorem of finite Abelian groups. Then in 1871 Dedekind produced a new edition of Dirichlet's number theory, with supplements he had written himself. These included the first version of his theory of ideals, though Dedekind did not yet know what to call it. He included it his Supplement X, called "On the Composition of Binary Quadratic Forms."

In the 1871 edition, only the first 43 pages of Supplement X (§§145–158) are about composition of quadratic forms. The next 75 pages (§§159–170) turn toward algebraic number fields, algebraic integers, and ideals. In Dedekind's next edition of Dirichlet, in 1879, the ideal theory is no longer buried in "On the Composition of Binary Quadratic Forms." It gets its own Supplement XI, called "On the Theory of Algebraic Integers," which now fills §§159–181 (nearly 200 pages). The old §§145–158 remain the same. Indeed, it seems likely that Dedekind kept §§145–158 mainly for the record, since §181 of the new Supplement on algebraic integers explains composition of quadratic forms in the language of quadratic integers (and in 17 pages).

In 1894 Dedekind wrote a third version, with §§145–158 still unchanged, but expanding "On the Theory of Algebraic Integers" to over 220 pages. Again, the final § of Supplement XI is the one from 1879 on composition of quadratic forms, slightly revised. It seems clear that, by this stage, Dedekind considered composition of forms to be a branch of the theory of algebraic integers, and of the related theory of rings and ideals.

In our next two chapters we will see how algebraic integers found a place in the theory of fields and rings, and how the problematic behavior of quadratic forms brought to light the concept of ideal. Thus, the problems arising from quadratic forms were eventually resolved by *abstraction*: choosing the right abstract concepts to make the problems clear. Nevertheless, at each stage, the choice of abstract concepts was guided by concrete examples from the world of algebraic integers.

It is noteworthy that Emmy Noether, who is responsible more than anyone else for the abstract theory of rings and ideals, urged her students to read all three versions of Dedekind's Supplements.

# 4

# Rings and Fields

## Preview

In previous chapters we have seen that the basic concepts of addition and multiplication on the ordinary integers quickly lead to difficult problems, even in finding integer solutions to quadratic and cubic equations.

At the same time, we have seen that the basic properties of addition and multiplication can be usefully encapsulated in the algebraic structures of rings and fields. These structures provide a setting for the generalized "integers" that illuminate the ordinary integers, and for the polynomials that define generalized integers.

We have also seen that, in a generalized integer setting, the property of unique prime factorization may fail. This creates a serious obstacle to the use of generalized integers, since success often depends on imitating arguments that are valid for ordinary integers.

Overcoming this obstacle requires a general investigation of rings and fields, with the ultimate aim of finding a substitute – **ideal prime factorization** – that succeeds in cases where ordinary prime factorization fails. The models for this investigation are the **algebraic number fields** and the rings of **algebraic integers** they contain.

The main aim of this chapter is to investigate algebraic number fields and to describe them as simply as possible. Once this is done, we define the **integers** of an algebraic number field $E$ and prove that they form a ring $\mathbb{Z}_E$. The properties of $\mathbb{Z}_E$ unfold alongside a deeper investigation of rings that begins in the next chapter.

## 4.1 Integers and Fractions

As we said in Sections 1.6 and 1.7, the system $\mathbb{Z}$ of ordinary integers satisfies the **ring axioms**

$$
\begin{aligned}
a + b &= b + a & ab &= ba & \text{(commutative laws)} \\
a + (b + c) &= (a + b) + c & a(bc) &= (ab)c & \text{(associative laws)} \\
a + (-a) &= 0 & & & \text{(inverse law)} \\
a + 0 &= a & a \cdot 1 &= a & \text{(identity laws)} \\
a(b + c) &= ab + ac & & & \text{(distributive law)}
\end{aligned}
$$

and the system $\mathbb{Q}$ of rational numbers satisfies the **field axioms**

$$
\begin{aligned}
a + b &= b + a & ab &= ba & \text{(commutative laws)} \\
a + (b + c) &= (a + b) + c & a(bc) &= (ab)c & \text{(associative laws)} \\
a + (-a) &= 0 & aa^{-1} &= 1 \text{ for } a \neq 0 & \text{(inverse law)} \\
a + 0 &= a & a \cdot 1 &= a & \text{(identity laws)} \\
a(b + c) &= ab + ac & & & \text{(distributive law)}
\end{aligned}
$$

However, it is worth reviewing the step that takes us from $\mathbb{Z}$ to $\mathbb{Q}$ – the construction of **fractions** – because we tend to forget what an important (and complicated!) step it is. By recalling this step in detail, we will be able to see how it generalizes to a larger class of rings.

The first thing to recall about fractions is that many fractions represent the same rational number. For example,

$$
\frac{1}{2} = \frac{-1}{-2} = \frac{2}{4} = \frac{-2}{-4} = \frac{3}{6} = \frac{-3}{-6} = \cdots .
$$

In general, if $a, b, a', b' \in \mathbb{Z}$ with $b, b' \neq 0$, then

$$
\frac{a}{b} = \frac{a'}{b'} \iff ab' = a'b.
$$

We may view a fraction $a/b$ abstractly as just an ordered pair $(a, b)$ of integers with $b \neq 0$, and call two pairs **equivalent** when they represent equal fractions. Then a **rational number** may be formally defined as an **equivalence class** of these pairs. Denoting equivalence by the symbol $\equiv$, we have

$$
(a, b) \equiv (a', b') \iff ab' = a'b. \tag{*}
$$

The relation $\equiv$ is an example of an **equivalence relation** because it has the properties of **reflexivity**, **symmetry**, and **transitivity**:

$$
\begin{aligned}
&(a, b) \equiv (a, b) & \text{(reflexivity)} \\
&(a, b) \equiv (a', b') \iff (a', b') \equiv (a, b) & \text{(symmetry)} \\
&(a, b) \equiv (a', b') \text{ and } (a', b') \equiv (a'', b'') \Rightarrow (a, b) \equiv (a'', b'') & \text{(transitivity)}
\end{aligned}
$$

The first two properties follow easily from the definition (*) and the corresponding properties of equality. To prove transitivity we need to prove the equivalent statement

$$ab' = a'b \text{ and } a'b'' = a''b' \Rightarrow ab'' = a''b.$$

Well,

$$ab' = a'b \text{ and } a'b'' = a''b'$$
$$\Rightarrow ab'b'' = a'bb'' \text{ and } a'bb'' = a''bb' \quad \text{(multiplying first by } b'', \text{ second by } b)$$
$$\Rightarrow ab'b'' = a''bb' \quad \text{(by transitivity of equality)}$$
$$\Rightarrow 0 = ab'b'' - a''bb' = b'(ab'' - a''b)$$
$$\Rightarrow 0 = ab'' - a''b \quad \text{(because } b' \neq 0)$$
$$\Rightarrow ab'' = a''b,$$

as required. In the penultimate line we have used the property that $\mathbb{Z}$ has **no zero divisors**: that is, if $x \neq 0$ and $y \neq 0$, then $xy \neq 0$.

The "no zero divisors" property is also needed for the second important fact about fractions: that they have a well-defined **sum** and **product**. Both the sum and product of $a/b$ and $a'/b'$ have the denominator $bb'$, so it is important that $bb' \neq 0$ when $b \neq 0$ and $b' \neq 0$. The definitions are

$$\frac{a}{b} + \frac{a'}{b'} = \frac{ab' + a'b}{bb'} \quad \text{and} \quad \frac{a}{b} \cdot \frac{a'}{b'} = \frac{aa'}{bb'},$$

Of course we all know that fractions behave like this, but we tend to forget how frustrating their behavior seemed at first encounter (especially the sum – why can't the answer be $\frac{a+a'}{b+b'}$?). Writing the definitions in terms of ordered pairs may help to reawaken the difficulty:

$$(a,b) + (a',b') = (ab' + a'b, bb') \quad \text{and} \quad (a,b) \cdot (a',b') = (aa',bb'). \quad (**)$$

It is *not obvious* that these definitions are independent of the pair chosen to represent a given equivalence class, and to check independence we have to go through maneuvers like those above in the proof of transitivity.

If $(a',b')$ is replaced by $(a'',b'')$ in the definition (**), then the pair for the sum, $(ab' + a'b, bb')$, is replaced by $(ab'' + a''b, bb'')$. According to the definition (*) of equivalence, the latter two pairs are equivalent just in case

$$(ab' + a'b)bb'' = (ab'' + a''b)bb'.$$

Subtracting $abb'b''$ from both sides leaves

$$a'b^2b'' = a''b^2b', \quad \text{or} \quad b^2(a'b'' - a''b') = 0,$$

which means, since $b \neq 0$ and there are no zero divisors, that

$$a'b'' - a''b' = 0, \quad \text{or} \quad a'b'' = a''b'.$$

This is precisely the condition for $(a', b')$ to be equivalent to $(a'', b'')$. There is a similar proof that the product of equivalence classes is well defined.

Finally, we can prove from the definitions of sum and product that the equivalence classes of fractions form a field, $\mathbb{Q}$. The proofs of the field properties are mainly routine applications of the definitions (**), combined with the known ring properties of $\mathbb{Z}$.

For example, to prove that addition in $\mathbb{Q}$ is commutative, consider

$$
\begin{aligned}
(a,b) + (a',b') &= (ab' + a'b, bb') && \text{(by definition of $+$)} \\
&= (a'b + ab', b'b) && \text{(by commutativity in $\mathbb{Z}$)} \\
&= (a',b') + (a,b) && \text{(by definition of $+$)}
\end{aligned}
$$

The other ring properties are proved similarly. Finally we have to prove that a nonzero $(a, b)$ has a multiplicative inverse. This is true because, when $a, b \neq 0$,

$$(a,b) \cdot (b,a) = (ab, ab) \equiv (1,1),$$

so $(b, a)$ is inverse to $(a, b)$. We saw in Section 1.7 that inverses are unique.

## Exercises

The idea that $\frac{a}{b} + \frac{a'}{b'}$ should be $\frac{a+a'}{b+b'}$ is false because $\frac{a+a'}{b+b'}$ fails to be independent of equivalence class representatives.

1. Give an example showing that $\frac{a+a'}{b+b'}$ has different values for different, but equivalent, pairs $(a', b')$.

The relation $a \equiv b \pmod{n}$ introduced in Sections 1.6 and 1.7 is an equivalence relation. We glossed over that fact at the time, but here is a check.

2. Explain why
   $a \equiv a \pmod{n}$
   $a \equiv b \pmod{n} \Leftrightarrow b \equiv a \pmod{n}$
   $a \equiv b \pmod{n}$ and $b \equiv c \pmod{n} \Rightarrow a \equiv c \pmod{n}$

This example also illustrates a property of **equivalence classes**, where the equivalence class of element $a$ consists of all elements equivalent to $a$. Namely, *any two equivalence classes are either identical or disjoint*. For the relation of congruence mod $n$ the equivalence classes are the *congruence classes* $[a]$.

3. Prove that if $[a] \neq [b]$, then $[a]$ and $[b]$ have no element in common.
4. Prove a corresponding result for any equivalence relation.

## 4.2 Domains and Fields of Fractions

We gave the painfully detailed treatment of fractions above because *exactly the same definitions and proofs* apply in any ring with no zero divisors.

**Definitions.** A **domain**[1] $D$ is a ring with no zero divisors; that is, if $a, b \in D$ are nonzero, then so is $ab$. The **fractions** of $D$ are the ordered pairs $(a, b)$, with $b \neq 0$ and **sum** and **product** defined by

$$(a, b) + (a', b') = (ab' + a'b, bb') \quad \text{and} \quad (a, b) \cdot (a', b') = (aa', bb'). \quad (**)$$

Their equivalence classes form the **field of fractions** of $D$, Frac(D).

Any field $F$ is an example of a domain, because if $a, b \in F$ with $ab = 0$ and $b \neq 0$, then $0 = abb^{-1} = a$. It is also clear that $F$ is its own field of fractions.

Other examples, which will be particularly important in this book, are rings of algebraic integers. We will give a general definition of these rings later, but a good example is the ring $\mathbb{Z}[\sqrt{2}] = \{a + b\sqrt{2}; a, b \in \mathbb{Z}\}$ introduced in Section 1.6.

$\mathbb{Z}[\sqrt{2}]$ has no zero divisors, because its elements are all real numbers, and the real numbers include no zero divisors. Thus, we can form fractions, which we will write in the usual way. The field of fractions of $\mathbb{Z}[\sqrt{2}]$ includes all the quotients of $a + b\sqrt{2}$, where $a, b \in \mathbb{Z}$, by $c \in \mathbb{Z}$, which are all the numbers of the form

$$p + q\sqrt{2} \quad \text{where} \quad p, q \in \mathbb{Q}. \quad (*)$$

Conversely, any quotient of members of $\mathbb{Z}[\sqrt{2}]$ is of this form. To see why, it suffices to check that multiplicative inverses of members of $\mathbb{Z}[\sqrt{2}]$ are of the form (*), because the product of numbers of the form (*) is obviously of the same form. As for the inverse, observe that

$$\frac{1}{a + b\sqrt{2}} = \frac{a - b\sqrt{2}}{(a - b\sqrt{2})(a + b\sqrt{2})}$$

$$= \frac{a - b\sqrt{2}}{a^2 - 2b^2} = \frac{a}{a^2 - 2b^2} - \frac{b}{a^2 - 2b^2}\sqrt{2},$$

which is of the form (*). Thus the field of fractions of $\mathbb{Z}[\sqrt{2}]$ is the field of numbers of the form (*), which we call $\mathbb{Q}(\sqrt{2})$. (The round brackets indicate that we allow the formation of quotients as well as sums, differences, and products.)

---

[1] Many algebra books call these rings *integral* domains, but the adjective "integral" occurs elsewhere in ring theory, as we will see, so it is best to avoid it here.

But there are many examples of rings that do have zero divisors, and hence cannot be extended to fields. An example is the ring of congruence classes mod 6, which came up in Section 1.7. In the congruence classes mod 6 we have $[2] \neq [0]$ and $[3] \neq [0]$ but $[2][3] = [6] = [0]$.

## Exercises

The fraction field of $\mathbb{Z}[\sqrt{2}]$ illustrates a property we will later observe for other fields $F$ of algebraic numbers: *Its elements are not merely quotients of algebraic integers, but quotients of the special form*:

$$\frac{\text{integer of } F}{\text{ordinary integer}}.$$

1. If $a, b, c, d \in \mathbb{Z}$, write $\frac{a+b\sqrt{2}}{c+d\sqrt{2}}$ in the form $\frac{l+m\sqrt{2}}{n}$, where $l, m, n \in \mathbb{Z}$.
2. Prove a similar result for quotients of members of $\mathbb{Z}[\sqrt{3}]$.
3. Find $\frac{1}{2^{1/3}-1}$, showing that it is actually in $\mathbb{Z}[2^{1/3}]$.

## 4.3 Polynomial Rings

Since its emergence as a separate mathematical discipline in the 16th century, algebra has involved polynomials – objects obtained from numbers and "unknowns" (denoted by $x, y, z, \ldots$ and other letters) by addition, subtraction, and multiplication. From the beginning it was recognized that polynomials obey the same rules of calculation as numbers; in other words, they have the *ring properties*, as we observed in Section 1.6.

Now we single out certain collections of polynomials, called **polynomial rings**, for special attention. Among them are:

1. $\mathbb{Z}[x]$, which consists of the polynomials in one unknown, $x$, with integer coefficients.
2. $\mathbb{Q}[x]$, which consists of the polynomials in one unknown, $x$, with rational coefficients.
3. $R[x]$, which consists of the polynomials in one unknown, $x$, with coefficients from a ring $R$. Or $F[x]$, which consists of the polynomials in one unknown, $x$, with coefficients from a field $F$.

In each case it is clear that the sum, difference, and product of two members of the ring is another member of the ring, and that the ring properties are satisfied.

Thus, each $R[x]$, where $R$ is a ring, is itself a ring. A typical polynomial in $R[x]$ takes the form

$$p(x) = a_n x^n + a_{n-1} x^{n-1} + \cdots + a_1 x + a_0, \quad \text{where } a_0, a_1, \ldots, a_n \in R.$$

Without loss of generality we can assume $a_n \neq 0$, in which case $p(x)$ is said to have **degree** $n$. (Unless $n = 0$, in which case we can have $p(x) = a_0$, which is of degree 0 if $a_0 \neq 0$ and by convention degree $-\infty$ if $a_0 = 0$.)

Since a ring $R$ gives a ring $R_1 = R[x_1]$, we can use $R_1$ in turn to form the ring $R_2 = R_1[x_2] = R[x_1, x_2]$, which consists of the two-variable polynomials with coefficients in $R$. Continuing in this way, we see that the $n$-variable polynomials with coefficients in $R$ form a ring $R[x_1, x_2, \ldots, x_n]$.

For suitable rings $R$, the polynomial ring $R[x]$ resembles $\mathbb{Z}$ in more than just the ring properties. If $R$ is a domain, then $R[x]$ is also a domain, so it has a field of fractions, denoted by $R(x)$. If $R$ is a field $F$, $F[x]$ also has a *Euclidean algorithm* and *unique prime factorization*. We observed this in the case of $F = \mathbb{Q}$ in Section 1.6, and the proof is exactly the same for coefficients in any field, since it involves only division and subtraction.

However, this is a good place to note that the gcd of two polynomials in $F[x]$ does not change when we allow divisors from $F'[x]$, for some field $F' \supseteq F$. Such divisors cannot arise, since the steps in the Euclidean algorithm do not produce any coefficients outside $F$. In particular, if we know the gcd by some other means – such as factorization in a larger $F'[x]$ – then we know that its coefficients are in $F$. This observation is used in Section 6.5.

### 4.3.1 Congruence Modulo a Polynomial

The analogy between $\mathbb{Z}$ and $F[x]$ goes even further. As we saw in Section 1.6, the step between the Euclidean algorithm and unique prime factorization involves an expression for the gcd of two polynomials $a(x)$ and $b(x)$:

$$\gcd(a(x), b(x)) = a(x)m(x) + b(x)n(x),$$

for some polynomials $m(x)$ and $n(x)$.

This too holds in $F[x]$ for any field $F$, because the division property in $F[x]$ (and hence the Euclidean algorithm) holds by exactly the argument given in Section 1.6. Another consequence of this form of the gcd is the existence of **inverses modulo a prime** polynomial $p(x)$.

**Definitions.** A nonconstant polynomial $p(x) \in R[x]$ is **prime** or **irreducible over** $R$ if it is not the product of lower-degree polynomials in $R[x]$. Polynomials $s(x)$ and $t(x)$ in $R[x]$ are **congruent mod** $p(x)$, written as

$$s(x) \equiv t(x) \pmod{p(x)},$$

if $p(x)$ divides $s(x) - t(x)$. The set of polynomials congruent to $s(x)$, mod $p(x)$, is called the **congruence class** of $s(x)$, written as $[s(x)]$.

**Existence of inverses.** *If $F$ is a field and $a(x)$ is not a multiple of the irreducible $p(x)$, then there is a polynomial $m(x) \in F[x]$ such that*

$$a(x)m(x) \equiv 1 \pmod{p(x)}.$$

*Proof.* It follows from the Euclidean algorithm that, for any $a(x), b(x) \in F[x]$ there are $m(x), n(x) \in F[x]$ such that

$$\gcd(a(x), b(x)) = a(x)m(x) + b(x)n(x).$$

In particular, if $a(x)$ is not a multiple of $p(x)$,

$$1 = \gcd(a(x), p(x)) = a(x)m(x) + p(x)n(x).$$

Thus, $p(x)$ divides $a(x)m(x) - 1$, which means

$$a(x)m(x) \equiv 1 \pmod{p(x)}. \qquad \square$$

**Corollary.** *The congruence classes mod an irreducible $p(x)$ form a field.*

*Proof.* The last congruence says that $[a(x)][m(x)] = [1]$, for any $a(x)$ not a multiple of $p(x)$; that is, for any *nonzero* congruence class $[a(x)]$. So each nonzero class $[a(x)]$ has a multiplicative inverse, $[m(x)]$. The other field properties of congruence classes (namely, the ring properties) are inherited from the ring properties of $F[x]$, exactly as for congruence classes in $\mathbb{Z}$ (Section 1.7). $\qquad \square$

## Exercises

It should be noted that the Euclidean algorithm for polynomials – like the one for integers but using division with remainder rather than subtraction – extends to an algorithm that computes the polynomials $m(x)$ and $n(x)$ such that

$$\gcd(a(x), b(x)) = a(x)m(x) + b(x)n(x).$$

We illustrate with an example from the exercises to Section 1.6.

1. Show

$$\gcd(x^3 - 2, x^2 - 1) = \gcd(x^2 - 1, x - 2) = \gcd(x - 2, 3) = 1,$$

via the following divisions with remainder:

$$x^3 - 2 = (x^2 - 1)x + x - 2 \qquad \text{(dividing } x^3 - 2 \text{ by } x^2 - 1)$$
$$x^2 - 1 = (x - 2)(x + 2) + 3 \qquad \text{(dividing } x^2 - 1 \text{ by } x - 2)$$

2. Deduce, by substituting $x - 2 = x^3 - 2 - (x^2 - 1)x$ from the first equation into the second, that

$$1 = (x^3 - 2)m(x) + (x^2 - 1)n(x),$$

where $m(x) = -(x + 2)/3$ and $n(x) = (1 + x)^2/3$.
3. Conclude that $(1 + x)^2/3$ is the inverse of $x^2 - 1$, modulo $x^3 - 2$.
4. Explain why $x^3 - 2$ is irreducible over $\mathbb{Q}$, but show that it factorizes in $\mathbb{R}[x]$.

## 4.4 Algebraic Number Fields

In the previous section the field of congruence classes of polynomials in $F[x]$, modulo an irreducible polynomial $p(x)$, was motivated partly by its analogy with the congruence classes of ordinary integers modulo a prime $p$. In both cases everything follows from the elementary idea of division with remainder.

In the case where $F = \mathbb{Q}$ there is a very natural interpretation of this field of congruence classes: as an **algebraic number field.**

**Definitions.** An **algebraic number** $\alpha$ is one[2] that satisfies an equation of the form

$$a_n x^n + a_{n-1} x^{n-1} + \cdots + a_1 x + a_0 = 0,$$

where $a_0, a_1, \ldots, a_n \in \mathbb{Q}$ and $a_n \neq 0$.

**Minimal polynomial.** *If $\alpha$ is an algebraic number then there is a unique polynomial, of minimal degree, that is satisfied by $\alpha$ and of the form*

$$p(x) = x^n + a_{n-1} x^{n-1} + \cdots + a_1 x + a_0 = 0, \quad \text{where } a_0, a_1, \ldots, a_{n-1} \in \mathbb{Q}.$$

*Also, any polynomial satisfied by $\alpha$ is a multiple of $p(x)$.*

---

[2] As is well known, the **fundamental theorem of algebra** (FTA) implies that any algebraic number lies in the field $\mathbb{C}$ of complex numbers. We will assume FTA later in this book. However, for our present purposes, it will suffice that the number is defined by a polynomial equation. Indeed, the existence of a solution to a polynomial equation will emerge *algebraically* from the ring $\mathbb{Q}[x]$ without appeal to analysis, unlike proofs of the fundamental theorem of algebra.

*Proof.* Clearly there is a $p(x)$ of minimal degree satisfied by $\alpha$. And we can get $p(x)$ in the above form by dividing through by the coefficient of the highest-degree term. Also, $p(x)$ is unique, because if there were a different minimal polynomial $p^*(x)$, also satisfied by $\alpha$, then $\alpha$ would satisfy $p(x) - p^*(x)$, which has *lower* degree.

Finally, suppose that $s(x)$ is any polynomial satisfied by $\alpha$. Then $s(x)$ necessarily has degree $\geq$ that of $p(x)$, so division with remainder by $p(x)$ gives

$$s(x) = q(x)p(x) + r(x), \quad \text{with degree } r(x) < \text{degree } p(x).$$

It follows that $\alpha$ also satisfies $r(x)$, and hence $r(x) = 0$ by minimality. Thus, any polynomial $s(x)$ satisfied by $\alpha$ is a multiple of $p(x)$. □

**Corollary.** *The minimal polynomial is irreducible over* $\mathbb{Q}$.

*Proof.* If $p(x)$ is a product $q(x)r(x)$ of polynomials in $\mathbb{Q}[x]$ of lower degree, then $\alpha$ satisfies one of $q(x) = 0$ or $r(x) = 0$, contrary to minimality. □

Now, despite the fact that we know *nothing* about the solutions of the equation $p(x) = 0$, we can construct a field in which this equation has a solution simply by saying "let $p(x) = 0$." This is the magic of algebra. More formally, we consider the congruence classes of $\mathbb{Q}[x]$ modulo $p(x)$. As we saw in the previous section, these congruence classes form a field in which (obviously) $p(x) = 0$. Moreover, this field is **isomorphic** to the fraction field $\mathbb{Q}(\alpha)$ of $\mathbb{Q}[\alpha]$, for any number $\alpha$ that satisfies $p(x) = 0$.

**Definition.** A field $F$ is **isomorphic to** a field $F'$, written $F \cong F'$, if there is a one-to-one map $\psi : F \to F'$, onto $F'$, such that

$$\psi(\beta + \gamma) = \psi(\beta) + \psi(\gamma), \quad \psi(\beta \cdot \gamma) = \psi(\beta) \cdot \psi(\gamma).$$

Such a map $\psi$ is called a **field isomorphism**.

**Solution field construction.** *The field of congruence classes of* $\mathbb{Q}[x]$ *modulo an irreducible polynomial* $p(x)$ *is isomorphic to* $\mathbb{Q}[\alpha]$, *for any solution* $\alpha$ *of the equation* $p(x) = 0$. *In particular,* $\mathbb{Q}[\alpha]$ *is a field, and hence equal to* $\mathbb{Q}(\alpha)$.

*Proof.* Since $p(x)$ is irreducible, its congruence classes form a field by the corollary in the previous section. In this field, the congruence class $[x]$ is a solution of $p(x) = 0$. In fact, the correspondence $[x] \mapsto \alpha$ defines an isomorphism from the congruence classes mod $p(x)$ to $\mathbb{Q}(\alpha)$, which is equal to its own fraction field. We unfold these facts in the following steps.

1. The map $[x] \mapsto \alpha$ extends to a map

$$\psi : \{\text{congruence classes mod } p(x)\} \to \mathbb{Q}[\alpha]$$

by setting $\psi([s(x)]) = s(\alpha)$ for any polynomial $s(x) \in \mathbb{Q}[x]$. This map is well defined because

$$[s(x)] = [t(x)] \Rightarrow p(x) \text{ divides } s(x) - t(x) \Rightarrow s(\alpha) = t(\alpha).$$

2. The map $\psi$ preserves sums and products because

$$\begin{aligned}
\psi([s(x)] + [t(x)]) = \psi([s(x) + t(x)]) \quad &\text{(by congruence class sum)} \\
= s(\alpha) + t(\alpha) \quad &\text{(by definition of } \psi) \\
= \psi([s(x)]) + \psi([t(x)]) \quad &\text{(by definition of } \psi)
\end{aligned}$$

Similarly for products – just replace the $+$ sign with the $\cdot$ sign.

3. The map $\psi : \{\text{congruence classes mod } p(x)\} \to \mathbb{Q}[\alpha]$ is onto $\mathbb{Q}[\alpha]$, because any $s(\alpha) \in \mathbb{Q}[\alpha]$ equals $\psi([s(x)])$. Thus, $\psi$ is an isomorphism onto $\mathbb{Q}[\alpha]$ and, since the congruence classes form a field, so does $\mathbb{Q}[\alpha]$. (Alternatively, we can see that $\mathbb{Q}[\alpha]$ includes the inverse of each nonzero $s(\alpha)$; namely $t(\alpha)$, where $[t(x)]$ is the inverse of the nonzero congruence class $[s(x)]$.) □

The field $\mathbb{Q}(\alpha) = \mathbb{Q}[\sqrt{2}]$ is a good illustration of this theorem. We already know that the domain

$$\mathbb{Q}[\sqrt{2}] = \{a + b\sqrt{2}; a, b \in \mathbb{Q}\}$$

equals its own fraction field. The theorem shows that $[x] \mapsto \sqrt{2}$ defines an isomorphism from the congruence classes of $\mathbb{Q}[x]$ modulo $x^2 - 2$ to $\mathbb{Q}[\sqrt{2}]$. In fact there are *two* such isomorphisms: $\psi^+$ sending $[x]$ to $\sqrt{2}$ and $\psi^-$ sending $[x]$ to $-\sqrt{2}$, since both $\sqrt{2}$ and $-\sqrt{2}$ are roots of $x^2 - 2 = 0$.

**Notation.** The field of congruence classes of $\mathbb{Q}[x]$ modulo an irreducible polynomial $p(x)$ is denoted by $\mathbb{Q}[x]/(p(x))$. (This notation is an instance of a more general notation – for the quotient of a ring by a principal ideal – that will be explained in Section 5.3. We already used a simple case of it in Section 3.6.)

### 4.4.1 Conjugates

Different solutions $\alpha$ and $\alpha'$ of the same equation $p(x) = 0$, where $p(x)$ is an irreducible polynomial, are called **conjugates** of each other. For example, $\sqrt{2}$

and $-\sqrt{2}$ are conjugate solutions of the equation $x^2 - 2 = 0$. We have already used special cases of this terminology in Chapters 1, 2, and 3.

The above proof that $\mathbb{Q}(\alpha)$ is isomorphic to the field $\mathbb{Q}[x]/(p(x))$ of congruence classes of $\mathbb{Q}[x]$ modulo $p(x)$ applies equally well to $\mathbb{Q}(\alpha')$, since $\alpha'$ also satisfies the equation $p(x) = 0$. Thus both $\mathbb{Q}(\alpha)$ and $\mathbb{Q}(\alpha')$ are isomorphic to the field of congruence classes, and hence *they are isomorphic to each other*.

Indeed, the proof shows that if $\alpha_1, \ldots, \alpha_n$ are the roots of $p(x) = 0$, then the map $\sigma_i : \alpha_1 \mapsto [x] \mapsto \alpha_i$ extends to an isomorphism $\sigma_i : \mathbb{Q}[\alpha_1] \to \mathbb{Q}[\alpha_i]$. Thus, when the minimal polynomial has $n$ roots, there are $n$ isomorphisms $[x] \mapsto \alpha_i$ of the field of congruence classes into a field containing all the roots of $p(x)$ (such as $\mathbb{C}$, by the fundamental theorem of algebra) .

## Exercises

The previous sections show that it is fairly easy to handle algebraic numbers arising from quadratic equations, such as $a + b\sqrt{2}$ where $a, b \in \mathbb{Q}$. For example, it is quite easy to find the inverse of $a + b\sqrt{2}$. The results of the present section allow us, in principle, to find inverses of algebraic numbers of arbitrary degree, by finding inverses of polynomials with the help of the Euclidean algorithm, as in the previous exercise set. Here is an example arising from a cubic equation.

1. Explain why $\mathbb{Q}(2^{1/3}) \cong \mathbb{Q}[x]/(x^3 - 2)$
2. Deduce that the inverse of $2^{2/3} - 1$ corresponds to the inverse of $x^2 - 1$, modulo $x^3 - 2$.
3. Hence calculate the inverse of $2^{2/3} - 1$ from exercise 4 of Section 1.6, expressing your answer in the form $a + 2^{1/3}b + 2^{2/3}c$ for some $a, b, c \in \mathbb{Q}$.
4. Find the inverse of $2^{2/3} + 1$ in the form $a + 2^{1/3}b + 2^{2/3}c$ for some $a, b, c \in \mathbb{Q}$.

## 4.5 Field Extensions

When a field $F$ is a subfield of a larger field $E$, we call $E$ an **extension field** of $F$, and we sometimes write $F \subseteq E$ or $E \supseteq F$ with the understanding that $F$ is not merely a subset of $E$ but also has the same sums and products. The properties of an extension field $E \supseteq F$ are often found from knowledge of the properties of $F$ together with knowledge of the properties of $E$ "relative to" or (as algebraists say) **over** $F$.

An example of an extension is the field $\mathbb{Q}(\sqrt{2}) \supseteq \mathbb{Q}$. Since $\sqrt{2}$ is irrational, $\mathbb{Q}(\sqrt{2})$ is a proper extension of $\mathbb{Q}$. By writing

$$\mathbb{Q}(\sqrt{2}) = \{a + b\sqrt{2} : a, b \in \mathbb{Q}\}$$

we can see one of the important "relative" properties of $\mathbb{Q}(\sqrt{2})$. It has **dimension 2 over** $\mathbb{Q}$ in the sense of linear algebra; namely:

- Every element of $\mathbb{Q}(\sqrt{2})$ is a linear combination, $a + b\sqrt{2}$, of the two elements $1, \sqrt{2}$.
- The latter two elements are **linearly independent over** $\mathbb{Q}$. That is, if the linear combination $a + b\sqrt{2} = 0$ with $a, b \in \mathbb{Q}$, then $a = b = 0$ (otherwise we would have $\sqrt{2} = -a/b$, which is rational).

**Definitions.** If $E \supseteq F$ is a field extension, then a **basis** for $E$ over $F$ is a set of elements $e_1, \ldots, e_k$ of $E$ such that

- The basis elements **span** $E$ over $F$; that is, each $e \in E$ has the form

$$e = f_1 e_1 + \cdots + f_k e_k \text{ for some } f_1, \ldots, f_k \in F.$$

- The basis elements are **linearly independent over** $F$; that is, if

$$0 = f_1 e_1 + \cdots + f_k e_k \text{ for some } f_1, \ldots, f_k \in F,$$

then $f_1 = \cdots = f_k = 0$.

$E$ is said to be **of finite dimension over** $F$ if it has a finite basis over $F$.

The elements of a field extension $E \supseteq F$ of finite dimension are called **algebraic over** $F$ because of the theorem below. We also say that $E$ itself is algebraic over $F$, or is an **algebraic extension** of $F$. The proof assumes the fact (hopefully familiar, but see Chapter 6 for a proof) that if $E$ has dimension $n$ over $F$, then any set of $n + 1$ elements of $E$ is linearly dependent over $F$.

**Finite-dimensional extensions.** *If $E$ is of finite dimension $n$ over $F$, then each $e \in E$ satisfies an equation of the form*

$$x^m + f_{m-1} x^{m-1} + \cdots + f_1 x + f_0 = 0,$$
$$\text{where } f_0, f_1, \ldots, f_{m-1} \in F \text{ and } m \leq n.$$

*Proof.* If $e \in E$, then $1, e, \ldots, e^n$ are $n + 1$ elements of $E$, and hence they are linearly dependent over $F$. That is, they satisfy an equation

$$f_n e^n + f_{n-1} e^{n-1} + \cdots + f_1 e + f_0 = 0,$$

where $f_0, f_1, \ldots, f_n \in F$ are not all zero. Obviously, $f_0$ cannot be the only nonzero coefficient. If $e^m$ is the largest power with nonzero coefficient, $f_m$, then we can divide through by $f_m$ and obtain an equation for $e$ of the form

$$x^m + f_{m-1}x^{m-1} + \cdots + f_1 x + f_0 = 0,$$

where $f_0, f_1, \ldots, f_{m-1} \in F$ and $m \leq n$.    $\square$

**Corollary.** *If $F \supseteq \mathbb{Q}$ is an extension of dimension $n$, then all members of $F$ are algebraic numbers of degree $\leq n$.*

## 4.5.1 The Extension $\mathbb{Q}(\alpha)$ of $\mathbb{Q}$

The most common field extension is the field $\mathbb{Q}(\alpha)$ obtained from $\mathbb{Q}$ by **adjoining** an algebraic number $\alpha$. Intuitively, this field consists of all numbers obtainable from $\alpha$ and rational numbers by the operations of $+, -, \times$, and $\div$. More formally, if $p(x)$ is the minimal polynomial for $\alpha$, then $\mathbb{Q}(\alpha)$ is the field we called $\mathbb{Q}[x]/(p(x))$ in the previous section.

The formal construction has the advantage of showing us a basis for $\mathbb{Q}(\alpha)$.

**Basis for $\mathbb{Q}(\alpha)$.** *If the minimal polynomial $p(x)$ for $\alpha$ has degree $n$, then the elements $1, \alpha, \alpha^2, \ldots, \alpha^{n-1}$ form a basis for $\mathbb{Q}(\alpha)$.*

*Proof.* The elements $1, \alpha, \alpha^2, \ldots, \alpha^{n-1}$ of $\mathbb{Q}(\alpha)$ correspond to the congruence classes $1, [x], [x]^2, \ldots, [x]^{n-1}$ of $\mathbb{Q}[x]$ modulo $p(x)$, so it suffices to show that these congruence classes span and are linearly independent over $\mathbb{Q}$.

To show that they span, suppose that

$$p(x) = x^n + a_{n-1}x^{n-1} + \cdots + a_1 x + a_0, \quad \text{where } a_0, a_1, \ldots, a_{n-1} \in \mathbb{Q}.$$

It follows that

$$x^n = -a_{n-1}x^{n-1} - \cdots - a_1 x - a_0,$$

so the class $[x]^n$ is a linear combination of the classes $1, [x], [x]^2, \ldots, [x]^{n-1}$. Next, if we multiply the later equation by $x$, we get

$$x^{n+1} = -a_{n-1}x^n - \cdots - a_1 x^2 - a_0 x,$$

showing that the class $[x]^{n+1}$ is a linear combination of the class $[x]^n$ and the classes $1, [x], [x]^2, \ldots, [x]^{n-1}$, so it is a combination of $1, [x], [x]^2, \ldots, [x]^{n-1}$ alone. Further multiplication by $x$ shows that the same is true of $[x]^{n+1}, [x]^{n+3}$, and so on.

Thus any polynomial in $[x]$ is a linear combination of $1, [x], [x]^2, \ldots, [x]^{n-1}$.

To show that $1, [x], [x]^2, \ldots, [x]^{n-1}$ are linearly independent, suppose on the contrary that

$$b_{n-1}[x]^{n-1} + \cdots + b_1[x] + b_0 = 0,$$

for some $b_0, b_1, \ldots, b_{n-1} \in \mathbb{Q}$, not all zero.

This means that $\alpha$ satisfies a polynomial of degree $\leq n - 1$, contrary to the minimality of $p(x)$.                                                                  □

This theorem is not as specialized as it looks, because in fact *any* extension field $E \supseteq \mathbb{Q}$ of finite dimension is of the form $\mathbb{Q}(\alpha)$. The number $\alpha$ is called a **primitive element** for $E$, and we prove its existence in Section 6.5.

## Exercises

The idea of a basis for one field $E$ over another field $F$ will be explored more thoroughly in Chapter 6, particularly in Section 6.2. However, we can prepare for the general discussion by looking at the example of $F = \mathbb{Q}(\sqrt{2})$ and its extension $E$ obtained by adjoining the element $\sqrt{3}$.

1. Show that $\sqrt{3} \notin \mathbb{Q}(\sqrt{2})$ by supposing

$$\sqrt{3} = a + b\sqrt{2}, \quad \text{where } a, b \in \mathbb{Q},$$

and finding a rational expression for $\sqrt{2}$.

Now any field containing $F = \mathbb{Q}(\sqrt{2})$ and $\sqrt{3}$ includes, at a minimum, the numbers $\alpha + \beta\sqrt{3}$ for $\alpha, \beta \in \mathbb{Q}(\sqrt{2})$. It is also clear that the sum and product of such numbers is another number of the same form.

2. Show that $\frac{1}{\alpha + \beta\sqrt{3}}$ is also of the form $\gamma + \delta\sqrt{3}$, where $\gamma, \delta \in F$, and hence that $E = \{\alpha + \beta\sqrt{3} : \alpha, \beta \in F\}$ is the smallest field containing $F$ and $\sqrt{3}$.
3. Explain why 1 and $\sqrt{3}$ form a basis for $E$ over $F$.

Finally, let us switch to viewing $E$ as a field over $\mathbb{Q}$.

4. Show that the numbers $\alpha + \beta\sqrt{3}$ for $\alpha, \beta \in F$ are precisely the numbers $a + b\sqrt{2} + c\sqrt{3} + d\sqrt{6}$ for $a, b, c, d \in \mathbb{Q}$.
5. By writing $a + b\sqrt{2} + c\sqrt{3} + d\sqrt{6}$ as

$$(a + b\sqrt{2})1$$
$$+ (c + d\sqrt{2})\sqrt{3}$$

and using exercise 3, show that $a + b\sqrt{2} + c\sqrt{3} + d\sqrt{6} = 0$ if and only if $a = b = c = d = 0$.
6. Conclude that $1, \sqrt{2}, \sqrt{3}, \sqrt{6}$ is a basis for $E$ over $\mathbb{Q}$.

In the exercises to Section 6.5 we will investigate the field $E$ further, showing that $\sqrt{2} + \sqrt{3}$ is a primitive element for $E$ and finding its minimal polynomial. Since these results require no further theory, the reader may wish to attempt them now.

## 4.6 The Integers of an Algebraic Number Field

The preceding sections have shown that fields are somewhat easier to handle than rings. In particular, they afford a nice agreement between the concepts of "dimension" and "degree." In Chapter 6 we will view this agreement in the setting of **vector spaces**, which are based on fields.

Since we have a clear and simple concept of algebraic number field – namely, a finite-dimensional extension of $\mathbb{Q}$ – it seems a good idea to place rings of **algebraic integers** inside algebraic number fields.

**Definitions.** An **algebraic integer** is a solution of an equation

$$x^n + a_{n-1}x^{n-1} + \cdots + a_1 x + a_0 = 0, \quad \text{where } a_0, \ldots, a_{n-1} \in \mathbb{Z}.$$

The algebraic integers in an algebraic number field $E$ are called the **integers of** the field $E$.

As an example, it follows from these definitions that the roots of $x^3 - 1 = 0$ are algebraic integers. In fact they are integers of the field $\mathbb{Q}(\sqrt{-3})$, since

$$x^3 - 1 = (x - 1)(x^2 + x + 1)$$

and the roots of $x^2 + x + 1 = 0$ are $\frac{-1 \pm \sqrt{-3}}{2}$. Thus, perhaps surprisingly, the integers of the field $\mathbb{Q}(\sqrt{-3})$ are not all of the form $a + b\sqrt{-3}$ with $a, b \in \mathbb{Z}$.

Allowing roots of all the polynomials $x^n + a_{n-1}x^{n-1} + \cdots + a_1 x + a_0$ (the so-called monic polynomials) to be "integers" begins to make sense when we observe that they behave as integers should, in the following sense.

**Sum, difference, and product of algebraic integers.** *If $\alpha$ and $\beta$ are algebraic integers, then so are $\alpha \pm \beta$ and $\alpha\beta$.*

*Proof.* Suppose that $\alpha$ satisfies an equation $\alpha^k + a_{k-1}\alpha^{k-1} + \cdots + a_1\alpha + a_0 = 0$. This gives

$$\alpha^k = -a_{k-1}\alpha^{k-1} - \cdots - a_1\alpha - a_0,$$
$$\alpha^{k+1} = -a_{k-1}\alpha^k - \cdots - a_1\alpha^2 - a_0\alpha,$$
$$\vdots$$

and so on. Thus, $\alpha^k$ is a linear combination of $1, \alpha, \ldots, \alpha^{k-1}$ with integer coefficients. It follows in turn that so too are $\alpha^{k+1}, \alpha^{k+2}, \ldots$ and hence every polynomial in $\alpha$ with integer coefficients.

Similarly, if $\beta$ satisfies a monic polynomial equation of degree $l$, then any polynomial in $\beta$ is a linear combination of $1, \beta, \ldots, \beta^{l-1}$ with integer coefficients. Therefore, *any polynomial in $\alpha$ and $\beta$ is a linear combination of terms $\alpha^i \beta^j$, with $0 \le i \le k-1$ and $0 \le j \le l-1$, with integer coefficients.*

So, if we denote the $kl$ products $\alpha^i \beta^j$ by $\omega_1, \ldots, \omega_{kl}$ we can write each polynomial $\omega$ in $\alpha$ and $\beta$ (such as $\alpha \pm \beta$ or $\alpha\beta$) in the form

$$\omega = n_1 \omega_1 + \cdots + n_{kl} \omega_{kl} \quad \text{where} \quad n_1, \ldots, n_{kl} \in \mathbb{Z}. \qquad (*)$$

From this we get $kl$ equations in the $kl$ "unknowns" $\omega_m$ by multiplying $(*)$ by $\omega_1, \ldots, \omega_{kl}$ and rewriting each right-hand side as a linear combination of the $\omega_m$ with integer coefficients:

$$\omega\omega_1 = n_1' \omega_1 + \cdots + n_{kl}' \omega_{kl}$$
$$\omega\omega_2 = n_1'' \omega_1 + \cdots + n_{kl}'' \omega_{kl}$$
$$\vdots$$
$$\omega\omega_{kl} = n_1^{(kl)} \omega_1 + \cdots + n_{kl}^{(kl)} \omega_{kl}.$$

These are $kl$ homogeneous equations in the $kl$ unknowns $\omega_m$, with a nonzero solution, so their determinant[3] must be zero. That is,

$$\det \begin{pmatrix} n_1' - \omega & n_2' & \cdots & n_{kl}' \\ n_1'' & n_2'' - \omega & \cdots & n_{kl}'' \\ \vdots & & & \vdots \\ n_1^{(kl)} & n_2^{(kl)} & \cdots & n_{kl}^{(kl)} - \omega \end{pmatrix} = 0.$$

The determinant is a polynomial in $\omega$ with integer coefficients, whose highest power $\omega^{kl}$ has coefficient $\pm 1$. Thus we get a monic polynomial equation for $\omega$, showing that $\omega = \alpha + \beta, \alpha - \beta$ or $\alpha\beta$ is an algebraic integer.     □

**Corollary.** *The integers of an algebraic number field form a ring.*

*Proof.* If $\alpha$ and $\beta$ are integers of an algebraic number field $E$, then $\alpha \pm \beta$, $\alpha\beta \in E$ since $E$ is a field. And these elements are algebraic integers by the theorem above. Also, the sums, differences, and products of integers in $E$ inherit the ring properties from $E$.     □

---

[3] In this chapter and the next, we assume some general properties of determinants. Readers probably know these properties from a linear algebra course, but we prove them from scratch in Chapter 7, so as to provide a single foundation for several results.

**Notation.** The ring of integers of a field $E \supseteq \mathbb{Q}$ of finite degree is called $\mathbb{Z}_E$.

One reason for placing algebraic integer rings in algebraic number fields, and hence bounding their degree, is that if $\alpha$ is an algebraic integer, then so is $\sqrt{\alpha}$. If we put no bound on degree, each algebraic integer has a "factorization"

$$\alpha = \sqrt{\alpha}\,\sqrt{\alpha}.$$

This rules out any chance of primes or prime factorization, which was the reason for using algebraic integers in the first place. Thus, we need to bound degree to have any hope of prime factorization. It is still not easy to achieve, as we will see, but there is hope in the concept of "ideal numbers," which is the starting point of our next chapter.

## Exercises

The simplest illustration of the concept of integer in an algebraic number field $E$ is where $E = \mathbb{Q}$. If this concept of "integer" makes sense, then the ring of integers of $\mathbb{Q}$ should be $\mathbb{Z}$. To confirm this, suppose that $x \in \mathbb{Q}$ satisfies the minimal equation

$$x^k + a_{k-1}x^{k-1} + \cdots + a_1 x + a_0 = 0, \quad \text{where } a_0, \ldots, a_k \in \mathbb{Z} \quad (*)$$

1. Suppose that $x = m/n$ satisfies (*), where $\gcd(m,n) = 1$ and $n > 1$, so $x \notin \mathbb{Z}$. Deduce that

$$m^k = -a_{k-1}n - \cdots - a_1 n^{k-1} - a_0 n^k,$$

and explain why this is a contradiction.

With the definition of algebraic integer we are now in a position to prove the result mentioned in the exercises to Section 4.2, that an algebraic number is the quotient of an algebraic integer by an ordinary integer (a result we will see again in Section 8.2).

2. Suppose that $\beta \in E$, an algebraic number field of finite degree, and that $\beta$ satisfies the polynomial

$$a_m x^m + a_{m-1}x^{m-1} + \cdots + a_1 x + a_0 = 0, \quad \text{where } a_0, \ldots, a_m \in \mathbb{Z}.$$

Then $\alpha = a_m \beta \in E$ also. Why?
3. By multiplying the above equation by $a_m^{m-1}$, explain why $\alpha$ satisfies the *monic* equation

$$\alpha^m + a_{m-1}\alpha^{m-1} + \cdots + a_1 a_m^{m-2}\alpha + a_0 a_m^{m-1} = 0.$$

4. Deduce that $\alpha$ is an integer of the field $E$, and hence that each $\beta \in E$ is of the form (integer of $E$)/(ordinary integer).
5. Explain why this shows that $E = \text{Frac}(\mathbb{Z}_E)$.

Finally, let us consider why each algebraic integer has a factorization in the ring of all algebraic integers.

6. If $\alpha$ is an algebraic integer, so is $\sqrt{\alpha}$. Why? Is $\sqrt[3]{\alpha}$ also an algebraic integer?

## 4.7 An Equivalent Definition of Algebraic Integer

The definition of algebraic integer given in the previous section is suitable for most purposes in this book, but it has one drawback: it is hard to tell when a given algebraic number is *not* an algebraic integer. For example, the number $\alpha = \sqrt{3}/\sqrt{2}$ is an algebraic number that satisfies the minimal monic polynomial equation

$$x^2 - \frac{3}{2} = 0,$$

which does not have integer coefficients. This could mean that $\alpha$ is not an algebraic integer. But how do we know there is no *other* monic polynomial equation for $\alpha$ with integer coefficients?

If there were such a polynomial, $x^2 - 3/2$ would divide it by the minimal polynomial theorem in Section 4.4. This contradicts the following result, which comes from Gauss (1801), article 42.

**Gauss's lemma.** *If a monic polynomial $f(x) \in \mathbb{Z}[x]$ factorizes into $g(x)h(x)$, where $g(x)$ and $h(x)$ are monic polynomials in $\mathbb{Q}[x]$, then in fact $g(x)$ and $h(x)$ are in $\mathbb{Z}[x]$.*

*Proof.* Following an idea of (Dedekind, 1871, p. 466), we introduce the **content** $\text{cont}(F)$ of a polynomial $F(x) \in \mathbb{Z}[x]$, which is the gcd of the coefficients of $F(x)$. We first prove that if $G(x), H(x) \in \mathbb{Z}[x]$, then

$$\text{cont}(GH) = \text{cont}(G)\text{cont}(H).$$

By first dividing each of $G(x)$ and $H(x)$ by the gcd of their coefficients, we can suppose that $\text{cont}(G) = \text{cont}(H) = 1$, so it remains to proves that $\text{cont}(GH) = 1$.

Suppose on the contrary that a prime number $p$ divides $\text{cont}(GH)$, so $p$ divides each coefficient of $G(x)H(x)$, where, say,

$$G(x) = b_r x^r + \cdots + b_1 x + b_0,$$
$$H(x) = c_s x^s + \cdots + c_1 x + c_0.$$

Our assumption $\text{cont}(G) = \text{cont}(H) = 1$ means there is a least $i$ such that $p$ does not divide $b_i$ and a least $j$ such that $p$ does not divide $c_j$. But then $p$ does not divide the coefficient of $x^{i+j}$ in $G(x)H(x)$, namely,

$$b_0 c_{i+j} + b_1 c_{i+j-1} + \cdots + b_{i-1} c_{j+1} + b_i c_j + b_{i+1} c_{j-1} + \cdots + b_{i+j} c_0,$$

since $p$ divides each term *except* $b_i c_j$. This contradiction shows $\text{cont}(GH) = 1$.

Now we return to the monic $f(x) \in \mathbb{Z}[x]$ that equals $g(x)h(x)$ for monic $g(x), h(x) \in \mathbb{Q}[x]$. We multiply each of $g(x)$ and $h(x)$ by an integer $m$ so that both $G(x) = mg(x)$ and $H(x) = mh(x)$ are in $\mathbb{Z}[x]$. Then, on the one hand,

$$\text{cont}(GH) = \text{cont}(m^2 f(x)) = m^2, \quad \text{because } f(x) \text{ is monic and in } \mathbb{Z}[x].$$

On the other hand, by the multiplicative property of content just proved,

$$m^2 = \text{cont}(G)\text{cont}(H).$$

Now $\text{cont}(G) = \text{cont}(mg) \leq m$ because $m = $ coefficient of $x^r$ in $G(x)$, and similarly $\text{cont}(H) = \text{cont}(mh) \leq m$. Hence, in fact, $\text{cont}(G) = \text{cont}(H) = m$ and so $\text{cont}(g) = \text{cont}(h) = 1$. That is, $g(x), h(x) \in \mathbb{Z}[x]$. $\square$

**Corollary.** *If $\alpha$ satisfies a monic polynomial in $\mathbb{Z}[x]$, then the minimal monic polynomial $g(x)$ for $\alpha$ in $\mathbb{Q}[x]$ is in $\mathbb{Z}[x]$.*

*Proof.* If $\alpha$ satisfies the monic polynomial $f(x) \in \mathbb{Z}[x]$ then the minimal monic polynomial $g(x)$ for $\alpha$ in $\mathbb{Q}[x]$ divides $f(x)$ by the minimal polynomial theorem of Section 4.4. Then $g(x) \in \mathbb{Z}[x]$ by Gauss's lemma. $\square$

Thus, an equivalent to the definition of algebraic integer given in the previous section is the following: An algebraic integer is an algebraic number whose *minimal* polynomial has integer coefficients.

## Exercises

1. Show that $\frac{\sqrt{3}}{\sqrt{2}}$ is not an algebraic integer, but that $\frac{1+\sqrt{3}}{\sqrt{2}}$ is.

Gauss's lemma is the key to the most useful test for irreducibility of polynomials in $\mathbb{Q}[x]$, called the **Eisenstein irreducibility criterion** after its appearance in Eisenstein (1850) (though it was actually found by Schönemann a few years earlier). The criterion states that *if*

$$f(x) = x^n + a_{n-1} x^{n-1} + \cdots + a_1 x + a_0 \in \mathbb{Z}[x]$$

Figure 4.1 Gotthold Eisenstein (1823–1852). Licensed under Creative Commons Attribution-ShareAlike 4.0 International License.

*and $p$ is a prime that divides each $a_i$ but $p^2$ does not divide $a_0$, then $f(x)$ is irreducible in $\mathbb{Q}[x]$.*

2. Suppose that $f(x)$ satisfies the criterion for a prime $p$ but that $g(x)$ and $h(x)$ are monic polynomials such that $f(x) = g(x)h(x)$. Deduce that $g(x)$ and $h(x)$ are in $\mathbb{Z}[x]$.
3. Now let $\overline{f}(x), \overline{q}(x), \overline{h}(x)$ be the polynomials resulting from $f(x), g(x), h(x)$, respectively, by replacing each coefficient by its congruence class mod $p$. Explain why $\overline{f}(x) = \overline{q}(x)\overline{h}(x)$.
4. Also, $\overline{f}(x) = x^n$ (why?). And unique factorization into irreducibles holds for polynomials with coefficients in $\mathbb{Z}/p\mathbb{Z}$ (why?). Deduce that $\overline{g}(x) = x^s$ and $\overline{h}(x) = x^t$ for some positive integers $s$ and $t$.
5. Deduce from exercise 4 that all but the leading coefficients of $g(x)$ and $h(x)$ are divisible by $p$, so $p^2$ divides $a_0$, which is a contradiction.

A famous application of the Eisenstein criterion is to prove irreducibility of the **cyclotomic[4] polynomial** $f(x) = x^{p-1} + x^{p-2} + \cdots + x + 1$, for prime values $p$. This polynomial is the nontrivial factor of $x^p - 1$, whose factorization is a special case of the one in Section 1.8.

---

[4] The word comes from the Greek for "circle dividing," and it is used here because the roots of the polynomial are $p$ equally spaced points on the unit circle in the plane of complex numbers. We saw examples of circle division in the exercises to Section 2.3.

6. Explain why $f(x) = \frac{x^p - 1}{x - 1}$ is irreducible if and only if $f(y + 1) = \frac{(y+1)^p - 1}{y}$ is irreducible.

7. Use the binomial theorem to show that

$$f(y + 1) = y^{p-1} + py^{p-2} + \frac{p(p - 1)}{2!}y^{p-3} + \cdots + p.$$

Also explain why all coefficients except the first are divisible by $p$.

8. Deduce, by Eisenstein's criterion, that $f(y + 1)$ is irreducible, and hence so is $f(x)$.

## 4.8 Discussion

Do you believe in irrational numbers? It depends on how much you need them. Analysts need not just individual irrational numbers like $\sqrt{2}$ or $\pi$, but also the infinite *totality* of real numbers, $\mathbb{R}$, and the totality of complex numbers $\mathbb{C}$, so most analysts have no difficulty believing in irrational numbers.

In algebra the need for infinite objects is less, and some eminent algebraists, such as Kronecker, have tried to avoid them. Of course this is not entirely possible, since the set $\mathbb{N}$ of natural numbers is infinite. But it is, as the Greeks used to say, only *potentially* infinite. One can avoid thinking of $\mathbb{N}$ as a completed whole by viewing it instead as an endless process, producing $0, 1, 2, 3, \ldots$ by repeatedly adding 1. Kronecker was willing to accept potentially infinite collections such as $\mathbb{N}$ but not *actually* infinite collections like $\mathbb{R}$ and $\mathbb{C}$.

This makes the so-called **fundamental theorem of algebra** (FTA) a tricky issue for algebraists. As it is normally stated, the FTA says that each polynomial equation $p(x) = 0$ has a root in $\mathbb{C}$. So the very statement of the theorem assumes the existence of the actual infinity $\mathbb{C}$. Also, this is the most convenient statement of FTA, because it admits an easy proof by standard theorems about continuous functions. But it is embarrassing that a supposed theorem of algebra is actually a theorem of analysis.

Kronecker (1887) proposed to replace FTA by what he called the "fundamental theorem of general arithmetic." His idea has not been embraced by mathematicians in general, but it led to the simplest and best description of algebraic number fields, which we used in Section 4.4. Let us now revisit the FTA, in order to better appreciate Kronecker's contribution.

### 4.8.1 The Fundamental Theorem of Algebra

The FTA was anticipated by mathematicians for centuries before it was actually proved. Hope for such a theorem was raised by the 16th-century

Figure 4.2 Niels Henrik Abel (1802–1829) and Evariste Galois (1811–1832).
Public domain and courtesy of the Bibliotèque nationale de France, respectively.

solution of cubic equations, mentioned in Section 1.9. This breakthrough was
quickly followed by the solution of quartic (degree 4) equations by Cardano's
student Ferrari, so it seemed reasonable to expect similar, if more complicated,
solutions for equations of degree 5, 6, and so on.

What the 16th-century mathematicians could not know was that equations
of degrees 2, 3, and 4 are special. They admit solutions by specific *formulas*
such as the quadratic formula

$$x = \frac{-b \pm \sqrt{b^2 - 4ac}}{2a} \quad \text{when} \quad ax^2 + bx + c = 0,$$

and the Cardano formula for the cubic given in Section 1.9. Formulas of this
kind, involving the coefficients of the equation and the operations $+, -, \times, \div$,
and $n$th roots, are *too simple* to solve even the general equation of degree 5.
This remarkable result was proved by Abel (1826a), and it was explained in
terms of group theory by the famous theory of Galois a few years later.

But before this happened, Gauss had realized that it was unnecessary to
seek formulas for solving polynomial equations, and better merely to prove
that *solutions exist*. Between 1799 and 1816 Gauss gave several proofs of FTA
that depend on properties of continuous functions and hence, ultimately, on
the properties of the infinite set $\mathbb{R}$. The simplest property that suffices to prove
FTA is the so-called **intermediate value theorem**, which states: if $f(x)$ is
continuous for $a \leq x \leq b$ and $f(a) < 0 < f(b)$, then $f(c) = 0$ for some $c$
between $a$ and $b$.

It follows from the intermediate value theorem that *any odd-degree polynomial equation $p(x) = 0$ has a real root*. This is because we can suppose
without loss of generality that the leading term of $p(x)$ is $x^{2m+1}$, in which

case $p(a) < 0$ for large negative $a$ and $p(b) > 0$ for large positive $b$. So, since any polynomial is continuous, it follows from the intermediate value theorem that $p(c) = 0$ for some $c$ between $a$ and $b$.

This argument is the nonalgebraic ingredient in the proof of FTA given by Gauss (1816). The remainder of the proof, while complicated, is a purely algebraic reduction of a general polynomial equation to one of odd degree (and quadratic equations, which of course we know how to solve). But the intermediate value theorem is not replacable by algebra—at least not the kind of algebra that Kronecker believed in. The existence of the intermediate value $c$ depends on the **completeness** of $\mathbb{R}$: the fact that $\mathbb{R}$ has no holes or gaps. Obvious though completeness may seem, it implies that $\mathbb{R}$ is an actual infinity.

The problematic nature of $\mathbb{R}$ had been a headache for mathematicians ever since the Pythagoreans discovered that $\sqrt{2}$ is irrational. The most subtle and difficult part of Euclid's *Elements*, Book V, is about getting a round the difficulties posed by irrational quantities. But the depth of the problem did not become clear until Cantor (1874) proved that $\mathbb{R}$ is **uncountable**. That is, it is impossible to arrange the set of real numbers the way we can arrange the natural numbers (and also the rational numbers, and even the algebraic numbers) as

first number, second number, third number, ...

This discovery, which we discuss further in Section 5.7, showed that $\mathbb{R}$ cannot be merely a "potential" infinity. It is probably no coincidence that Kronecker was vehemently opposed to Cantor's work, and that his campaign to avoid infinity began as Cantor's discoveries became known.

### 4.8.2 Algebraic Numbers

The surprising subtlety of FTA, particularly its entanglement with infinity, raised doubts about the meaning and existence of algebraic numbers in the minds of Kronecker and others. If there is no simple and uniform formula for solving polynomial equations, can there be a simple and uniform way of dealing with algebraic numbers?

Yes! Kronecker realized that the way to handle any algebraic number is by the solution field construction in Section 4.4. Namely, instead of constructing the solution $\alpha$ of an irreducible polynomial equation $p(x) = 0$, construct the *solution field* $\mathbb{Q}(\alpha)$ as the field of congruence classes of $\mathbb{Q}[x]$ mod $p(x)$. In this field, $\alpha$ is represented by the congruence class $[x]$, so all we want to know about $\alpha$, algebraically, can be found by finite calculations with rational polynomials.

This idea illustrates an approach that is common in modern mathematics: The behavior of an individual object (such as $\alpha$) may sometimes be best seen by embedding it in a larger object (such as $\mathbb{Q}[x]/(p(x))$) which has "more structure" (in this case, the structure of a field).

Before Kronecker and Dedekind, and before there were doubts about the nature of $\mathbb{R}$, Cauchy (1847) used a solution field to give meaning to the (then still dubious) concept of $\sqrt{-1}$. He embedded it as $[x]$ in the field of real polynomials in $x$ modulo $x^2 + 1$. This field is isomorphic to the field $\mathbb{C}$ of complex numbers. Thus, if one is willing to use $\mathbb{R}$, one should also be willing to use $\mathbb{C}$.

### 4.8.3 Algebraic Integers

The philosophy of "understanding through embedding in a larger structure" is implemented all the more when it comes to understanding algebraic integers. An algebraic integer is, first of all, an algebraic number, and hence it comes to us embedded in a field of algebraic numbers. But *which* members of the field should be regarded as integers?

In the beginning, Euler (1770) made the lucky guess that numbers of the form $a + b\sqrt{-2}$, where $a, b \in \mathbb{Z}$, should be regarded as "integers." In the 1840s, Kummer also guessed correctly that the numbers in $\mathbb{Z}[\zeta_n]$, where

$$\zeta_n = \cos \frac{2\pi}{n} + i \sin \frac{2\pi}{n},$$

should be considered "integers" (called **cyclotomic integers**, because they divide the unit circle into $n$ equal parts), and indeed they are the integers of the field $\mathbb{Q}(\zeta_n)$.

However, it would be a mistake to suppose that the numbers of the form $a + b\sqrt{-3}$, where $a, b \in \mathbb{Z}$, are all the integers of the field $\mathbb{Q}(\sqrt{-3})$. In fact, $\frac{-1+\sqrt{-3}}{2}$ should also be considered an integer of this field. We have already seen merit in this viewpoint in the exercises to Section 2.9, where it was shown that unique prime factorization fails in $\mathbb{Z}[\sqrt{-3}]$, but is regained in $\mathbb{Z}\left[\frac{-1+\sqrt{-3}}{2}\right]$. But why should $\frac{-1+\sqrt{-3}}{2}$ be considered an integer? The number $\frac{-1+\sqrt{-3}}{2}$ is the same as $\zeta_3$, so this brings us back to the question of why $\zeta_n$ should be considered an integer.

The answer has to do with the equation satisfied by $\zeta_n$, namely $x^n = 1$. This is a *monic* polynomial equation with ordinary integer coefficients, and Eisenstein (1850) proved that the sum, difference, and product of numbers satisfying such equations is another number of the same kind. We proved this theorem in Section 4.6. So the numbers satisfying monic polynomials form a *ring*, which (from our viewpoint today) is a good sign. Somehow realizing that such numbers are exactly what we want integers to be, (Dedekind, 1871, §160) *defined* algebraic integers to be the numbers satisfying monic polynomials in $\mathbb{Z}[x]$.

Later we will see that monic polynomials are the "right" concept to define "integers" in more general settings.

# 5

# Ideals

## Preview

As we know from $\mathbb{Z}$ and $\mathbb{Q}$, divisibility in rings is more interesting than in fields, because one nonzero element of a ring usually does not divide another. This exposes the concepts of primes and unique prime factorization. To study primes and factorization in $\mathbb{Z}$, we introduced the "microscope" of algebraic integer rings $\mathbb{Z}_E$, which can see inside ordinary integers by factoring them into algebraic integers.

But unique prime factorization does not always hold in $\mathbb{Z}_E$, and this reveals what makes rings more complicated than fields: the presence of substructures called **ideals**. Ideals are so called because they realize (among other things) objects once called "ideal numbers." These "numbers" were conjectured to exist by Kummer, in the hope of restoring unique prime factorization in the rings $\mathbb{Z}_E$. In effect, Kummer was looking for a "stronger microscope" than $\mathbb{Z}_E$ – one that can see inside the algebraic integers themselves.

Dedekind brought Kummer's "ideal numbers" into reality as ideals. He then showed, with appropriate concepts of prime ideal and factorization, that **unique prime ideal factorization** holds in the algebraic integer rings $\mathbb{Z}_E$.

In this chapter we introduce some ideals and the basic concepts around them: **congruence** modulo an ideal, **homomorphisms**, and the **quotient** $R/I$ of a ring $R$ by an ideal. We also begin searching for a description of the rings that admit unique prime ideal factorization, by investigating some ideals in the rings $\mathbb{Z}_E$.

This leads to the concept of **Noetherian** rings, which is the first step towards describing the right kind of ring for unique prime ideal factorization.

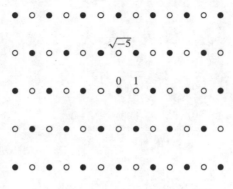

Figure 5.1 Multiples of $\gcd\left(2, 1 + \sqrt{-5}\right)$.

## 5.1 "Ideal Numbers"

We saw in Section 3.2 that prime factorization seems to fail in $\mathbb{Z}\left[\sqrt{-5}\right]$, since 6 has the two factorizations

$$6 = 2 \cdot 3 = \left(1 + \sqrt{-5}\right)\left(1 - \sqrt{-5}\right)$$

into numbers that are "prime" in the sense that they are not products of numbers with smaller norm in $\mathbb{Z}\left[\sqrt{-5}\right]$. In such a situation, Kummer believed that the numbers $2, 3, 1 + \sqrt{-5}, 1 - \sqrt{-5}$ could be further split into "ideal factors" so that the "ideal factorizations" of $2 \cdot 3$ and $\left(1 + \sqrt{-5}\right)\left(1 - \sqrt{-5}\right)$ are identical.

Pursuing this train of thought further, the ideal factors ought to include the "greatest common divisor" of 2 and $1 + \sqrt{-5}$. And what might the gcd be? Our study of gcd in $\mathbb{Z}$ gives a clue. There we found (Section 1.2) that

$$\gcd(a, b) = ma + nb, \quad \text{for some } m, n \in \mathbb{Z}.$$

This means $\gcd(a, b)$ is the *least* positive integer of the form $ma + nb$, since $\gcd(a, b)$ divides all numbers of the form $ma + nb$. It follows that the numbers $ma + nb$ are *all the integer multiples* of $\gcd(a, b)$.

We cannot find an actual member of $\mathbb{Z}\left[\sqrt{-5}\right]$ to serve as $\gcd\left(2, 1 + \sqrt{-5}\right)$, but we can find a *subset* of $\mathbb{Z}\left[\sqrt{-5}\right]$ to serve as the "multiples of $\gcd\left(2, 1 + \sqrt{-5}\right)$," namely

$$I = \left\{ 2\mu + \left(1 + \sqrt{-5}\right)\nu : \mu, \nu \in \mathbb{Z}\left[\sqrt{-5}\right] \right\}.$$

This set, which in fact is $\{2m + \left(1 + \sqrt{-5}\right)n : m, n \in \mathbb{Z}\}$, is shown amongst all the members of $\mathbb{Z}\left[\sqrt{-5}\right]$ as the set of black dots in Figure 5.1.

Figure 5.1 confirms that $I$ is *not* the set of multiples of any member of $\mathbb{Z}\left[\sqrt{-5}\right]$, because the multiples of any $\alpha \in \mathbb{Z}\left[\sqrt{-5}\right]$ form a lattice of

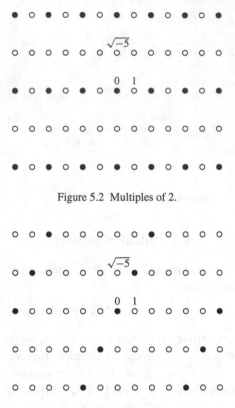

Figure 5.2 Multiples of 2.

Figure 5.3 Multiples of $1 + \sqrt{-5}$.

rectangles *of the same shape as* $\mathbb{Z}[\sqrt{-5}]$; namely, magnified by $|\alpha|$ and rotated by the angle of $\alpha$. The examples in Figures 5.2 and 5.3, whose black dots are the multiples of 2 and $1 + \sqrt{-5}$, respectively, show such magnified and rotated copies of $\mathbb{Z}[\sqrt{-5}]$. The black set in Figure 5.1, however, clearly does not consist of rectangles.

Thus, the "ideal number" $\gcd(2, 1 + \sqrt{-5})$ can be realized by its "set of multiples" $I$, which is a *subset* of $\mathbb{Z}[\sqrt{-5}]$, but not as a *member* of $\mathbb{Z}[\sqrt{-5}]$. This is not such a wild idea. We have already seen, in Section 1.7, that the natural way to view members of the finite field $\mathbb{F}_p$ is as *subsets* of $\mathbb{Z}$; namely, as congruence classes mod $p$. There are no members of $\mathbb{Z}$ that behave like the members of $\mathbb{F}_p$. It was Dedekind's idea to realize "ideal numbers" as certain sets, which he called **ideals**.

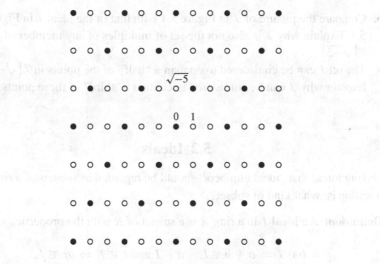

Figure 5.4 Multiples of gcd $(3, 1 + \sqrt{-5})$.

## Exercises

We stumbled on the "ideal number" $I$ in the course of trying to explain the two "prime" factorizations of 6 in $\mathbb{Z}[\sqrt{-5}]$, $2 \cdot 3$ and $(1 + \sqrt{-5})(1 - \sqrt{-5})$. Namely, $I$ represents the "greatest common divisor" of 2 and $1 + \sqrt{-5}$. This prompts us to investigate the "greatest common divisor" $J$ of 3 and $1 + \sqrt{-5}$, and to compare it with $I$.

1.  By writing $\mu = a + b\sqrt{-5}$ and $\nu = c + d\sqrt{-5}$, where $a, b, c, d \in \mathbb{Z}$, check that

    $$\left\{ 2\mu + (1 + \sqrt{-5})\nu : \mu, \nu \in \mathbb{Z}[\sqrt{-5}] \right\}$$
    $$= \left\{ 2m + (1 + \sqrt{-5})n : m, n \in \mathbb{Z} \right\}.$$

2.  Similarly, check that

    $$\left\{ 3\mu + (1 + \sqrt{-5})\nu : \mu, \nu \in \mathbb{Z}[\sqrt{-5}] \right\}$$
    $$= \left\{ 3m + (1 + \sqrt{-5})n : m, n \in \mathbb{Z} \right\},$$

    so that $\{3m + (1 + \sqrt{-5})n : m, n \in \mathbb{Z}\}$ is also an "ideal number" $J$ of $\mathbb{Z}[\sqrt{-5}]$ (which we might call "the multiples of gcd $(3, 1 + \sqrt{-5})$").

3. Compare the picture of $J$ in Figure 5.4 with that of the ideal $I$ in Figure 5.1. Explain why $J$ is also not the set of multiples of any member of $\mathbb{Z}\left[\sqrt{-5}\right]$.

4. The set $I$ can be considered to contain a "half" of the points in $\mathbb{Z}\left[\sqrt{-5}\right]$. Explain why $J$ can be considered to contain a "third" of these points.

## 5.2  Ideals

Having found that "ideal numbers" should be regarded as subsets of a ring, the question is: what kind of subset?

**Definition.** An **ideal** $I$ in a ring $R$ is a subset of $R$ with the properties

$$a, b \in I \Rightarrow a + b \in I, \quad a \in I \text{ and } r \in R \Rightarrow ar \in I.$$

In other words, $I$ is a subset of $R$ closed under addition, and under multiplication by elements of $R$. It follows that $I$ is also closed under additive inverse, because $-1 \in R$ and hence $a(-1) = -a \in I$ for any $a \in I$. This implies in turn that $I$ is closed under differences. The set we used in the previous section to represent the "ideal number" $\gcd\left(2, 1 + \sqrt{-5}\right)$ is easily seen to be an ideal by this definition. We can also see that the "ideal numbers" of $\mathbb{Z}$ correspond to the ordinary integers.

**Ideals in $\mathbb{Z}$.** *Each ideal in $\mathbb{Z}$ has the form $\{na : n \in \mathbb{Z}\}$, for some $a \in \mathbb{Z}$.*

*Proof.* If $I \neq \{0\}$ let $a$ be the least positive member of $I$. Then $I$ includes all multiples $an$ for $n \in \mathbb{Z}$, by closure under sums and additive inverses. If $I$ also includes some $b \neq an$ for any $n \in \mathbb{Z}$, and if $am$ is the nearest member of $I$ less than $b$, then $I$ also includes $b - am$.

But $0 < b - am < a$ by the division property of $\mathbb{Z}$ (Section 1.1), so we have a contradiction.                                                                    □

An ideal of the form $\{ar : r \in R\}$ for some $a \in R$ is called the **principal ideal generated by** $a$, and is often written $(a)$. Domains such as $\mathbb{Z}$, whose ideals are all principal, are called **principal ideal domains**, or PIDs. We saw in the previous section that $\mathbb{Z}\left[\sqrt{-5}\right]$ is *not* a PID, and that one **nonprincipal ideal** in $\mathbb{Z}\left[\sqrt{-5}\right]$ is $\{2m + \left(1 + \sqrt{-5}\right)n : m, n \in \mathbb{Z}\}$.

The latter ideal has two generators, 2 and $1 + \sqrt{-5}$, and we will see later that each ideal in $\mathbb{Z}\left[\sqrt{-5}\right]$ has a finite number of generators (one or two). Rings in which each ideal is finitely generated are in a sense the next best thing to PIDs, and we study them further in Section 5.4.

Coming back to ideals in general, for any ideal $I$ in a ring $R$ we have a concept of **congruence modulo** $I$. This leads to congruence classes, which themselves form a ring when added and multiplied in the obvious way. The ring of congruence classes, like other rings or fields of congruence classes we have seen, is a **quotient** of the ring $R$.

**Definition and notation.** Elements $a$ and $b$ of a ring $R$ are called **congruent modulo an ideal** $I$ if $a - b \in I$. We write $a \equiv b \pmod{I}$.

**The ring of congruence classes.** *Congruence modulo $I$ is an equivalence relation, whose congruence classes $[a]$ for the $a \in R$ form a ring under addition and multiplication defined by*

$$[a] + [b] = [a + b], \quad [a][b] = [ab].$$

*Proof.* That congruence modulo $I$ is an equivalence relation follows easily from the closure properties of $I$:

- Each $a \equiv a \pmod{I}$ because $a - a = 0 \in I$.
- If $a \equiv b \pmod{I}$, then $a - b \in I$, hence also $b - a \in I$, so $b \equiv a \pmod{I}$.
- If $a \equiv b \pmod{I}$ and $b \equiv c \pmod{I}$, then $a - b, b - c \in I$. Hence, their sum $a - c \in I$, which means $a \equiv c \pmod{I}$.

Consequently, $R$ is partitioned into equivalence classes $[a]$, and we have to check that the operations $+$ and $\cdot$ are well defined on these equivalence classes. The proof is the same as that for congruence classes in $\mathbb{Z}$ (Section 1.7), which used only the ring properties of sum and product.

Finally, the $+$ and $\cdot$ operations on congruence classes inherit the ring properties from $R$, exactly as they did for $\mathbb{Z}$. $\square$

We call the ring of congruence classes modulo $I$ the **quotient** of $R$ by $I$, and denote it by $R/I$. The quotient ring is just part of a circle of ideas we investigate further in the next section.

### 5.2.1 PIDs and Unique Prime Factorization

Before we go more deeply into the nature of ideals – particularly nonprincipal ideals – we point out that PIDs always have unique prime factorization, so they have no need for "ideal numbers."

In a PID $R$ we call an element $p \in R$ a **prime** if all divisors of $p$ are of the form $u$ or $up$, where $u$ is a divisor of 1 (a "unit"). Divisors of an element $r$ *not* of the form $u$ or $ur$ will be called **proper divisors** of $r$. The proofs of existence and uniqueness of prime factorization are like those for $\mathbb{Z}$, except that we have no Euclidean algorithm and no measure of the "size" of members of $R$.

**Unique prime factorization in a PID.** *In a PID, each element has a prime factorization, which is unique up the order of the primes and unit factors.*

*Proof.* (Existence). Given a PID $R$, take any element $r_1 \in R$. If $r_1$ is a prime, we are done. If not, $r_1$ has a proper divisor $r_2$, and we repeat the argument with $r_2$. Continuing in this way, we obtain a sequence of elements of $R$,

$$r_1, \quad r_2, \quad r_3, \quad \ldots,$$

each of which is a proper divisor of the one before. This gives an increasing sequence of principal ideals (an **ascending chain**, as we will call it later),

$$(r_1) \subset (r_2) \subset (r_3) \subset \cdots,$$

whose union is also an ideal. Then, since $R$ is a PID, there is an $s_1 \in R$ such that

$$(r_1) \cup (r_2) \cup (r_3) \cup \cdots = (s_1).$$

It follows that $s_1 \in (r_n)$ for some $n$, so $r_n$ is a divisor of $s_1$. But no proper divisors of $s_1$ are in $(s_1)$, so in fact $(r_n) = (s_1)$ and hence the sequence of proper divisors $r_1, r_2, r_3, \ldots$ ends with $r_n$. Thus, $r_n = s_1$ has no proper divisors and hence is prime.

Having found one prime divisor $s_1$ of $r_1$, we similarly find a prime divisor $s_2$ of $r_1/s_1$, and so on. This process terminates for a similar reason – considering the ideals $(r_1) \subset (r_1/s_1) \subset (r_1/s_1 s_2) \subset \cdots$ – so we get a prime factorization of any element of $R$.

(Uniqueness). As usual, it suffices to prove the prime divisor property: if a prime $p$ divides a product $ab$, then $p$ divides $a$ or $p$ divides $b$. So we suppose that $p$ divides $ab$, but $p$ does not divide $a$. As always, the latter assumption means that units are the only common divisors of $a$ and $p$.

Now the set $\{ma + np : m, n \in R\}$ is an ideal, hence of the form $(c)$ since $R$ is a PID. But then $c$ divides both $a$ and $p$, so $c$ is a unit and $(c) = R$. In particular,

$$1 = ma + np \quad \text{for some } m, n \in R.$$

We are now back on familiar ground: multiply both sides of this equation by $b$,

$$b = mab + npb,$$

observe that $p$ divides both terms on the right, so $p$ divides $b$, as required.  □

## Exercises

1. Generalize the proof that $\mathbb{Z}$ is a PID to show that any Euclidean domain (Section 1.6) is a PID.
2. Adapt the proof above to show that any PID $R$ satisfies the **ascending chain condition**: if $I_1 \subseteq I_2 \subseteq I_3 \subseteq \cdots$ are ideals of a PID $R$, then the sequence $I_1, I_2, I_s, \ldots$ is eventually constant.
3. Give examples of ideals $I$ in the rings $\mathbb{Z}$ and $\mathbb{Q}[x]$ where $R/I$ is a field, and others where $R/I$ is not a field.

## 5.3 Quotients and Homomorphisms

The relation of congruence modulo an ideal is a characteristic of rings that they do not share with fields, since the only ideals of a field $F$ are $(0)$ and $F$ itself. We therefore introduce some language and terminology for discussing the notions around ideals and congruence in rings.

**Definitions and notation.** The ring of congruence classes modulo $I$ in $R$ is called the **quotient** of $R$ by $I$ and is denoted by $R/I$. The map

$$\psi: R \to R/I \quad \text{defined by} \quad \psi(a) = [a],$$

where $[a]$ is the congruence class[1] of $a$ mod $I$, is called the **canonical homomorphism** of $R$ onto $R/I$.

A **ring homomorphism** in general is a map of rings $f: R \to R'$ such that $f(a + b) = f(a) + f(b)$ and $f(a \cdot b) = f(a) \cdot f(b)$. Its **kernel** is $\{a \in R: f(a) = 0\}$.

The next theorem shows that any ring homomorphism is essentially the canonical homomorphism for an ideal $I$, with kernel $I$. Readers who know group homomorphisms and their kernels will notice the analogy between rings

---

[1] It would be more precise to write this congruence class as $[a]_I$, to show its dependence on $I$, but it is usually obvious which ideal defines the congruence.

and groups in the theorem below: just as the kernel of a group homomorphism is a normal subgroup, the kernel of a ring homomorphism is an ideal.

**Homomorphisms and ideals.** *If* $f: R \to R'$ *is a ring homomorphism onto* $R'$, *then* $I = \{a \in R: f(a) = 0\}$ *is an ideal and the elements of* $R'$ *correspond to the congruence classes of* $R$ *mod* $I$.

*Proof.* By definition of $I$,

$[a] = [b] \Leftrightarrow a - b \in I \Leftrightarrow f(a - b) = 0$

$$\Leftrightarrow f(a) - f(b) = 0 \quad \text{(because } f \text{ is a homomorphism)}$$

$$\Leftrightarrow f(a) = f(b).$$

Thus the congruence classes mod $I$ correspond to elements of $R'$.

The equations $f(a + b) = f(a) + f(b)$ and $f(ab) = f(a)f(b)$ are equivalent to

$$[a + b] = [a] + [b] \quad \text{and} \quad [ab] = [a] \cdot [b],$$

which we proved to hold in the previous section.                                      $\square$

Our results about congruence in $\mathbb{Z}$ and other rings can be, and often are, written in the language of homomorphisms, ideals, and quotients.

The congruence classes $[a]$ of integers mod $n$ are the congruence classes modulo the principal ideal $\{an: n \in \mathbb{Z}\}$, which we denote by $a\mathbb{Z}$. The ring of these classes is therefore the quotient ring $\mathbb{Z}/a\mathbb{Z}$. In particular, the finite field $\mathbb{F}_p$ equals $\mathbb{Z}/p\mathbb{Z}$. (These are the notations we foreshadowed in Section 3.6.)

We could also write the quotient $\mathbb{Z}/a\mathbb{Z}$ as $\mathbb{Z}/(a)$, since $a\mathbb{Z}$ is the principal ideal generated by $a$, which we write as $(a)$. An instance where the principal ideal notation is usual is the ring of congruence classes of polynomials in $\mathbb{Q}[x]$ modulo an irreducible polynomial $p(x)$, which we introduced in Section 4.4. It is usual to denote this ring by $\mathbb{Q}[x]/(p(x))$.

We will see in the next section that when the quotient of a ring by an ideal is a field, as is the case for $\mathbb{Z}/(p)$ and $\mathbb{Q}[x]/(p(x))$, then the ideal is **maximal**. That is, there is no larger ideal except the whole ring. To see why the ideal $p\mathbb{Z}$ is maximal in $\mathbb{Z}$, consider an ideal $I$ containing $p\mathbb{Z}$ and some integer $q$ not a multiple of $p$. Then there are integers $m$ and $n$ such that

$$1 = \gcd(p, q) = mp + nq,$$

and $mp + nq \in I$ because $I$ is closed under sums and differences. But then $I$ contains all integer multiples of 1, so $I = \mathbb{Z}$.

## Exercises

The congruence classes of $\mathbb{Z}[\sqrt{-5}]$ modulo the ideals $I$ and $J$ of the previous section provide interesting examples of quotient rings and maximality.

1. Explain why there are two congruence classes of $\mathbb{Z}[\sqrt{-5}]$ modulo $I$; namely, the classes [0] and [1].
2. Deduce that $\mathbb{Z}[\sqrt{-5}]/I$ is a ring with two elements.
3. Explain why this ring is the field $\mathbb{F}_2$.
4. Explain why there are three congruence classes of $\mathbb{Z}[\sqrt{-5}]$ modulo $J$; namely, the classes [0], [1], and [−1].
5. Explain why the ring $\mathbb{Z}[\sqrt{-5}]/J$ is the field $\mathbb{F}_3$ with three elements.

In the next section we will see that this means $I$ and $J$ are maximal ideals. However, in the case of $I$ and $J$, we can also prove maximality directly.

6. Show that any ideal of $R = \mathbb{Z}[\sqrt{-5}]$ containing $I$ and some $x \notin R$ is the whole of $R$.
7. Similarly, show that any ideal of $R$ containing $J$ and some $x \notin R$ is the whole of $R$.

## 5.4  Noetherian Rings

In a general ring the argument that maximality of an ideal is equivalent to the quotient being a field goes as follows.

**Characterization of a maximal ideal.** *An ideal $I$ in a ring $R$ is maximal if and only if $R/I$ is a field.*

*Proof.* By the following sequence of equivalences:

$I$ is maximal $\Leftrightarrow$ $R$ = (ideal containing $I$ and any $q \neq I$)

$\Leftrightarrow$ $1 \in$ (ideal containing $I$ and any $q \neq I$)

$\Leftrightarrow$ for any $q \notin I$ there is $r \in R$ with $rq \equiv 1 \pmod{I}$

$\Leftrightarrow$ for any $[q] \neq [0]$ there is $r \in R$ with $[r][q] = [1]$

$\Leftrightarrow$ any $[q] \neq [0]$ has an inverse, mod $I$

$\Leftrightarrow$ $R/I$ is a field.

because we have already proved in Section 5.2 that $R/I$ is a ring.                    □

This theorem enables us to see that the nonprincipal ideal in $\mathbb{Z}[\sqrt{-5}]$,

$$I = \left\{ 2m + (1 + \sqrt{-5})n : m,n \in \mathbb{Z}[\sqrt{-5}] \right\},$$

which we called the "multiples of $\gcd(2, 1 + \sqrt{-5})$" in Section 5.1, is a maximal ideal of $\mathbb{Z}[\sqrt{-5}]$. This is because it is clear from Figure 5.1 that there are just two congruence classes mod $I$ in $\mathbb{Z}[\sqrt{-5}]$: the class $[0]$, which is $I$ itself, and the class $[1]$, which is the complement of $I$. Therefore, the quotient $\mathbb{Z}[\sqrt{-5}]/I$ is a ring with two elements, which is necessarily the *field* $\mathbb{F}_2$, since the only nonzero element, $[1]$, is self-inverse.

Every ring in which $0 \neq 1$, called a *nonzero* ring, has a maximal ideal. The proof, due to Krull (1929), is rather subtle. It is easier to exhibit a ring containing a collection of ideals with no maximal member. For example, in the ring of *all* algebraic integers one has the increasing sequence of ideals

$$\{\text{multiples of } 2^{1/2}\} \subset \{\text{multiples of } 2^{1/4}\} \subset \{\text{multiples of } 2^{1/8}\} \subset \cdots$$

which obviously has no maximal member. Thus the ring of all algebraic integers has infinite ascending chains of ideals – for much the same reason that it does not have "primes," already noted in Section 4.6.

**Definition.** A ring $R$ is **Noetherian** if every nonempty set $\mathcal{C}$ of ideals of $R$ has a maximal member; that is, one not properly contained in any other member.

This concept is named after the great algebraist Emmy Noether. She introduced an equivalent defining property (the **ascending chain condition** below) in Noether (1921). The Noetherian property is vital for the rings we study in our pursuit of unique prime ideal factorization. Its importance is clearer when one sees some equivalent statements of it, which relate containment to finite generation. To distinguish clearly between containment relations, we write $A \subset B$ or $B \supset A$ for "$A$ is contained in but not equal to $B$," and $A \subseteq B$ or $B \supseteq A$ for "$A$ is contained in or equal to $B$."

**Equivalents of the Noetherian property.** *A ring $R$ is Noetherian if it has any of the following three properties:*

1. *Any nonempty set of ideals of $R$ has a maximal member.*
2. *Any nondecreasing sequence $I_1 \subseteq I_2 \subseteq I_3 \subseteq \cdots$ of ideals of $R$ eventually becomes constant. (Ascending chain condition, ACC.)*
3. *Any ideal $I$ of $R$ is finitely generated; that is, there are $e_1, \ldots, e_n \in I$ such that $I = \{r_1 e_1 + \cdots + r_n e_n : r_1, \ldots, r_n \in R\}$.*

*Proof.* We prove the equivalence of 1, 2, and 3 by showing $1 \Rightarrow 3 \Rightarrow 2 \Rightarrow 1$.

Figure 5.5 Emmy Noether (1882–1935) (public domain).

$1 \Rightarrow 3$. Let $I$ be any ideal of $R$ and consider the collection of finitely generated ideals $\subset I$. By 1, this collection has a maximal member, $M$. Since $M \subset I$, there is a $q \in I$ not in $M$. But then $q$ and the generators of $M$ define a finitely generated ideal $J \subseteq I$ with $J \supset M$. Hence, in fact $J = I$, so $I$ is finitely generated.

$3 \Rightarrow 2$. Let $I_1 \subseteq I_2 \subseteq I_3 \subseteq \cdots$ be a nondecreasing sequence of ideals. The union $I$ of this sequence is also an ideal, hence it has a finite set of generators $e_1, \ldots, e_n$ by 3. Each $e_i \in$ some $I_j$, hence *all* of $e_1, \ldots, e_n \in$ some $I_k$. But then the sequence $I_1 \subseteq I_2 \subseteq I_3 \subseteq \cdots$ is constant from $I_k$ onwards.

$2 \Rightarrow 1$. Suppose that $C$ is a set of ideals of $R$ with no maximal member. In other words, for any $I \in C$ there is an $I' \in C$ with $I \subset I'$.

Choose any $I_1 \in C$. Then $I_1$ is not maximal, so we can choose $I_2 \in C$ with $I_1 \subset I_2$. Likewise, $I_2$ is not maximal, so we can choose $I_3 \in C$ with $I_3 \subset I_3$. Continuing in this way, we obtain an infinite ascending chain of ideals, contrary to ACC. Therefore, every nonempty set of ideals in $R$ has a maximal member.                                                    □

It is common in algebra books to prove the existence of maximal elements with the help of a principle of set theory known as **Zorn's lemma**, which is equivalent to the **axiom of choice**. In the last step of the proof above we also used choice, because we were not able to *define* the sequence $I_1, I_2, I_3$, and so on. But this choice principle – involving a *sequence* of choices, each depending on the one before – is simpler and arguably more plausible, called **dependent choice**. The dependent choice principle is known to be weaker than the full axiom of choice, or Zorn's lemma, but like them, it is not provable from the other axioms of set theory.

### Exercises

Zorn's lemma is about a collection $C$ of sets, the containment relation $\subseteq$, and subcollections $\mathcal{B}$ that are *linearly ordered* by containment; that is, if $B_i, B_j$ are in $\mathcal{B}$, then either $B_i \subseteq B_j$ or $B_j \subseteq B_i$. The lemma states that $C$ includes a maximal element if each linearly ordered subcollection $\mathcal{B}$ has an upper bound $B$; that is, $B \supseteq$ each $B_i$ in $\mathcal{B}$.

One of the simplest cases where we really need Zorn's lemma is the proof of Krull's theorem that any ring containing nontrivial ideals contains a maximal ideal. Suppose $R$ is a ring containing ideals $I$ other than $(0)$ and $R$ itself. Let $\mathcal{I}$ be the collection of all such ideals in $R$. To apply Zorn's lemma, we have to show that any linearly ordered subset of $\mathcal{I}$ is contained in a member of $\mathcal{I}$.

1. If $\mathcal{B}$ is a linearly ordered subset of $\mathcal{I}$, prove that the union $B$ of all members of $\mathcal{B}$ is an ideal. Where did you use the linear ordering?

Obviously $B \neq (0)$, but why is $B \neq R$? Here is a hint.

2. Explain why no $I \in \mathcal{I}$ includes the element 1, and deduce that $B \in \mathcal{I}$.
3. Conclude, by Zorn's lemma, that $R$ contains a maximal ideal.

## 5.5 Noether and the Ascending Chain Condition

The definition of the Noetherian property in terms of maximality is *not* given in the founding paper on the subject, Noether (1921). In fact, Noether proves the equivalence of the other two statements of the condition, thereby avoiding obvious use of a maximal principle such as Zorn's lemma. Her starting point is the assumption that ideals are finitely generated, a condition inspired by the famous *finite basis theorem* of Hilbert (1890). In her first theorem (the "theorem of the finite chain") she proves that the finite generation of ideals implies the ascending chain condition as in the step 3 $\Rightarrow$ 2 in the proof above.

Then she notes without proof that finite generation of ideals follows from the ascending chain condition, so the latter can also be taken as a finiteness condition "in basis-free form." The proof is short and almost obvious, but it is worth stating explicitly, because it too involves dependent choice.

**ACC implies ideals are finitely generated.** *If $R$ is a ring satisfying ACC, then any ideal $I$ of $R$ is finitely generated.*

*Proof.* Suppose $I$ is an ideal of $R$ that is not finitely generated. Choose an element $e_1 \in I$. Then the ideal $I_1$ generated by $e_1$ is not all of $I$, so we can choose an $e_2 \in I$ not in $I_1$. Then the ideal $I_2$ generated by $e_1, e_2$ is also not all of $I$, so we can choose an $e_3 \in I$ that is not in $I_2$.

Continuing in this way, we obtain a strictly increasing sequence of ideals $I_1 \subset I_2 \subset I_3 \subset \cdots$, contrary to the ascending chain condition.

Hence, in fact $I$ is generated by some finite set of elements (a "basis").     □

### 5.5.1 Hilbert's Basis Theorem

The Hilbert basis theorem will not be used later in this book; nevertheless, it is worth giving a proof, because of the influence the theorem had on Emmy Noether and the formulation of ACC in particular. Hilbert's proof, in Hilbert (1890), implicitly used ACC. It also used dependent choice, in order to obtain an infinite sequence that was not obviously definable. This provoked the criticism that Hilbert was doing "not mathematics, but theology." Modern versions of the theorem mention the Noetherian assumption explicitly in both the statement and the proof, and they tend to downplay Hilbert's original goal – which was to show the existence of a finite "basis" (meaning generating set) for any ideal in certain polynomial rings.

**Hilbert's basis theorem.** *If $R$ is a Noetherian ring, then the ring $R[x]$ is also Noetherian.*

*Proof.* Suppose, for the sake of contradiction, that an ideal $I \subseteq R[x]$ is not finitely generated. Thus, for any polynomials $f_0, \ldots, f_n \in I$, the ideal

$$(f_0, \ldots, f_n) = \{r_0 f_0 + \cdots + r_n f_n : r_0, \ldots, r_n \in R\}$$

is not the whole of $I$, and hence there is a polynomial $f_{n+1}$ in the set difference $I - (f_0, \ldots, f_n)$.

Then, using dependent choice, we can choose a sequence of polynomials $f_0, f_1, f_2, \ldots \in I$ so that

$$f_0 \in I \quad \text{is of minimal degree,}$$
$$f_1 \in I - (f_0) \quad \text{is of minimal degree,}$$
$$f_2 \in I - (f_0, f_1) \quad \text{is of minimal degree,}$$

$$\vdots$$

Since we choose minimal degree at each step,

$$\deg(f_0) \le \deg(f_1) \le \deg(f_2) \le \cdots .$$

Now let $a_n$ be the coefficient of the highest power of $x$ in $f_n$, and consider the ideals in $R$:

$$(a_0) \subseteq (a_0, a_1) \subseteq (a_0, a_1, a_2) \subseteq \cdots .$$

Since $R$ is Noetherian, this ascending chain of ideals becomes constant, say at $(a_0, \ldots, a_N)$. Now $a_N \in (a_0, \ldots, a_{N-1})$, so

$$a_N = r_0 a_0 + \cdots + r_{N-1} a_{N-1} \quad \text{for some } r_0, \ldots, r_{N-1} \in R.$$

This leads us to consider the polynomial

$$g = r_0 x^{\deg(f_N) - \deg(f_0)} f_0 + \cdots + r_{N-1} x^{\deg(f_N) - \deg(f_{N-1})} f_{N-1},$$

whose highest-power term

$$(r_0 a_0 + \cdots + r_{N-1} a_{N-1}) x^N = a_N x^N$$

is the highest-power term in $f_N$. By its definition, $g \in (f_0, \ldots, f_{N-1})$. And also $f_N \notin (f_0, \ldots, f_{N-1})$, by definition of $f_N$, so $f_N - g \notin (f_0, \ldots, f_{N-1})$. But

$$f_N - g \in I - (f_0, \ldots, f_{N-1}) \quad \text{has degree less than } N = \deg(f_N),$$

which contradicts the minimality of $f_N$.                                    □

Hilbert was actually interested in finding a finite basis for polynomial ideals in many variables, but this follows from the above formulation of the theorem. If $R$ is Noetherian, then so is $R' = R[x_1]$, and hence so is $R'' = R'[x_2] = R[x_1, x_2]$, and so on.

## Exercises

Any ideal $K$ of $\mathbb{Z}[\sqrt{-5}]$ is finitely generated, so $\mathbb{Z}[\sqrt{-5}]$ is Noetherian. In fact $K$ is generated by (at most two) elements nearest to 0 in different directions. The proof assumes only that $K$ is closed under sums and differences.

1. Let $u$ be a nonzero member of $K$ of least absolute value. Show that all elements of $K$ on the line through 0 and $u$ are integer multiples $mu$ of $u$.
2. Next, let $v$ be the least nonzero member of $K$ not on the line through 0 and $u$. Show that the elements $mu + nv$ divide the plane into parallelograms with adjacent sides of length $|u|$ and $|v|$.

3. Thus any $w \in K$ not of the form $mu + nv$ lies in one of the parallelograms, either in the interior of the parallelogram or in the interior of an edge. Show that this leads to an element $x \in K$ such that $|x| < |u|$ or $|x| < |v|$, contrary to the definition of $u$ and $v$.

Hilbert's basis theorem shows one way – passing from $R$ to $R[x]$ – in which the Noetherian property passes from one ring to another. Another way is by forming a quotient. Namely, if $I$ is an ideal of a Noetherian ring $R$, then $R/I$ is also Noetherian. To prove this, use homomorphisms as in Section 5.3.

4. Given a Noetherian ring $R$, a homomorphism $f : R \to R'$ onto ring $R'$, and an ideal $I'$ of $R'$, show that $f^{-1}(I')$ is an ideal of $R$.
5. Show that a finite basis of the ideal $f^{-1}(I')$ gives a finite basis of $I'$.
6. Conclude that $R/I$ is Noetherian for any ideal $I$ of a Noetherian ring $R$.

This gives another, purely algebraic, proof that $\mathbb{Z}[\sqrt{-5}]$ is Noetherian.

7. Express $\mathbb{Z}[\sqrt{-5}]$ in the form $\mathbb{Z}[x]/I$.

## 5.6 Countable Sets

The step $2 \Rightarrow 1$ in proving the equivalents of the Noetherian condition is more interesting in the case of a ring $R$ with *countably* many ideals, where the axiom of choice is not needed. The rings $\mathbb{Z}_E$ have this property,[2] so it is worth saying more about countability.

A set $S$ is called **countable** if $S = \{x_0, x_1, x_2, x_3, \ldots\}$. In other words, the members of $S$ can be "listed" as $x_0, x_1, x_2, x_3 \ldots$, so each member of $S$ is $x_n$ for some natural number $n$. Examples of countable sets (with suitable listings) are

$$\mathbb{N} = \{0, 1, 2, \ldots\}$$
$$\mathbb{Z} = \{0, 1, -1, 2, -2, , 3, -3, \ldots\}$$
$$\mathbb{N} \times \mathbb{N} = \{(0,0), (0,1), (1,0), (0,2), (1,1), (2,0), (0,3), (1,2), (2,1), (3,0), \ldots\}.$$

In the latter case the ordered pairs $(m, n)$ are listed in groups: first those with $m + n = 0$, then those with $m + n = 1$, and so on. And within each group, pairs are listed in **lexicographic order**, or "dictionary order." This ordering device can also be used to list pairs, triples, and so on, enabling us to show countability of sets such as $\mathbb{Q}$, $\mathbb{Z}[\sqrt{2}]$, and indeed any algebraic number field or ring.

---

[2] As was remarked by (Noether, 1926, footnote 26), expanding on a very brief remark she made about the axiom of choice in Noether (1921).

It follows that an algebraic number ring $\mathbb{Z}_E$ has countably many ideals, since (as we will eventually see) these ideals are finitely generated and hence may be listed by listing the finite generating sets, using lexicographic order. Thus it suffices to prove $2 \Rightarrow 1$ in the case of a set $R$ and a countable collection $C$ of its subsets. The proof is simple and purely about set containment.

**Existence of a maximal element.** *Suppose $C$ is a countable collection of subsets $I$ of a set $R$ such that any nondecreasing sequence*

$$I_1 \subseteq I_2 \subseteq I_3 \subseteq \cdots$$

*eventually becomes constant. Then $C$ has a maximal element.*

*Proof.* Suppose on the contrary that $C$ is a countable collection of subsets of $R$, satisfying the ascending chain condition, but with no maximal element. Let $I_1'$ be any member of $C$. Then $I_1'$ is not maximal, so there is a *first* member $I_2'$ in the listing of $C$ such that $I_1' \subset I_2'$. Similarly, $I_2'$ is not maximal, so there is a first member $I_3'$ in the listing of $C$ such that $I_2' \subset I_3'$.

Continuing in this way gives an infinite chain $I_1' \subset I_2' \subset I_3' \subset \cdots$, contrary to the ascending chain condition. Hence, $C$ has a maximal element. $\square$

The crucial difference between this proof and the proof for an arbitrary collection $C$ is the ability to find a *first* member of $C$ satisfying a certain condition, and hence define the infinite ascending chain $I_1' \subset I_2' \subset I_3' \subset \cdots$. In the general case the axiom of choice allows us to *assume* an ordering of $C$ for which there is a "first" element satisfying any condition. Such an ordering is called a **well ordering**, and its existence is equivalent to the axiom of choice (and to Zorn's lemma). It is fair to say that the axiom of choice is a kind of magic that allows us to imitate the proof for countable sets.

## Exercises

The listing of ordered pairs $(m,n)$ can be given by an explicit quadratic function known as the *Cantor pairing function*, which corresponds to the visually natural way of ordering the pairs shown in Figure 5.6.

1. Taking the diagonals to be ordered from left to right, and the pairs on each diagonal from top to bottom, explain why $(m,n)$ is the $m$th element on the $(m+n)$th diagonal.
2. Deduce that the position of $(m,n)$ in the ordering of pairs is

$$m + (0 + 1 + 2 + \cdots + (m+n)) = m + \frac{(m+n)(m+n+1)}{2},$$

when $(0,0)$ is taken to be at position 0.

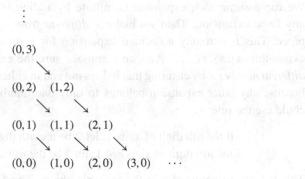

Figure 5.6 Ordering the pairs of natural numbers.

This proves that the function $P(m, n) = m + \frac{(m+n)(m+n+1)}{2}$ maps the set $\mathbb{N} \times \mathbb{N}$ of ordered pairs one-to-one onto $\mathbb{N}$.

3. Use $P$ to construct a one-to-one function from $\mathbb{N} \times \mathbb{N} \times \mathbb{N}$ to $\mathbb{N}$.

## 5.7 Discussion

Since we have singled out the axiom of choice and countability in the last two sections, we should not fail to mention what makes them pertinent: *the existence of uncountable sets*, proved first by Cantor (1874) and with greater impact by Cantor (1891). In particular, Cantor showed the uncountability of $\mathbb{R}$ by the following argument.

Suppose we have a countable set of real numbers $x_1, x_2, x_3, \ldots$ . We are going to show these are not all the real numbers by defining a real number $x$ different from each of $x_1, x_2, x_3, \ldots$ . This is easy to do if we suppose we have the decimal expansion for each $x_i$, say

$$x_1 = 0.\mathbf{1}1111\ldots$$
$$x_2 = 3.1\mathbf{4}159\ldots$$
$$x_3 = 1.41\mathbf{4}21\ldots$$
$$x_4 = 0.000\mathbf{0}0\ldots$$
$$x_5 = 1.2122\mathbf{1}\ldots$$
$$\vdots$$
$$x = 0.21112\ldots$$

We can assume each expansion is infinite by adding $0000\ldots$ at the end of any finite expansion. Then we make $x$ *different from $x_n$ in the $n$th decimal place*. This is certainly a decimal expansion for $x$ different from the given expansions $x_1, x_2, x_3, \ldots$. And we can make sure the expansion represents a *different number $x$* by ensuring that it does not end in either $0000\ldots$ or $9999\ldots$ because any other expansion belongs to only one number. For example, we could use the rule:

> if the $n$th digit of $x_n$ is 1, let 2 be the $n$th digit of $x$;
> if the $n$th digit of $x_n$ is *not* 1, let 1 be the $n$th digit of $x$.

This is how we obtained $x$ in the example above. The digits of $x_1, x_2, x_3, \ldots$ shown in bold are the digits where the rule makes $x$ different from $x_1, x_2, x_3, \ldots$ in turn. Because these digits lie along the diagonal, the method is called the **diagonal argument**.

This argument shows that no countable set includes all real numbers, and hence $\mathbb{R}$ is uncountable.

### 5.7.1 The Axiom of Choice

The axiom of choice asserts the possibility of "choosing" members of sets where there is no apparent rule for doing so. The precise statement is the following.

**Axiom of choice** (AC). For any set $X$ of nonempty sets $Y$, there is a function $f$ on $X$ with $f(Y) \in Y$ for each $Y \in X$. (The function $f$ is called a **choice function**.)

To many people, AC seems obvious, so they willingly accept all of its consequences. Indeed, when the sets $Y$ are all subsets of a countable set $Z$, the existence of a choice function is provable. We suppose $z_1, z_2, z_3, \ldots$ is a list of the members of $Z$, and let $f(Y)$ be the first member of $Y$ on this list.

But in the presence of uncountable sets, it is easy to give examples where no one can define a choice function. For example, let

$$X = \{\text{all nonempty sets } Y \subseteq \mathbb{R}\}.$$

There is no known rule for choosing a real number from each nonempty set of reals, and in fact the standard axioms of set theory cannot prove the existence of a choice function in this case. That is why AC is an *axiom*! It is an axiom about uncountable sets.

AC is a particularly remarkable axiom, not only because of its independence from the other axioms of set theory – proved by a combination of results due to Gödel (1938) and Cohen (1963) – but also because it has many natural-sounding equivalents. Thus, we might say that AC is the "right axiom" to prove these equivalent theorems. The first of them was the **well-ordering theorem**, first stated by Cantor (1883) as a "fundamental law of thought." The well-ordering theorem extends the well-ordering property of the natural numbers, mentioned in Section 1.1, by asserting that any *uncountable* set can be ordered in such a way that each of its nonempty subsets has a least member.

Zermelo (1904) introduced AC in order to prove the well-ordering theorem, and it is easy to show that well-ordering implies AC. Thus the well-ordering theorem is an equivalent of AC.

A second equivalent is **Zorn's lemma**, described in the exercises to Section 5.4. Zorn's lemma is actually due to Kuratowski (1922), but it was rediscovered by Zorn (1935), and it became the "algebraist's axiom of choice" when popularized by Bourbaki. Because Zorn's lemma asserts existence of maximal elements, it gives almost instant proofs that maximal algebraic objects exist. Two examples where existence of the algebraic object is actually *equivalent* to AC are:

- **Existence of a maximal ideal in any nonzero ring.** This was proved by Krull (1929), assuming the well-ordering theorem. Hodges (1979) proved that existence of maximal ideals implies AC.
- **Existence of a vector space basis.** This was proved (essentially) by Hamel (1905), though he was interested in a special case we will discuss in the exercises to Section 6.1. Blass (1984) proved that the existence of vector space bases implies AC.

However, there are cases where the weaker **axiom of dependent choice** suffices, as in equivalence proof for the three statements of the Noetherian property. Dependent choice is weaker than AC because it is known not to imply some consequences of AC. A famous question that distinguishes the two forms of choice concerns *nonmeasurable sets*: Vitali (1905) showed that AC implies there are nonmeasurable subsets of $\mathbb{R}$, but Solovay (1970) showed it is consistent with dependent choice for all subsets of $\mathbb{R}$ to be Lebesgue measurable.

### 5.7.2 Axioms of Infinity for Countable Algebra

As we showed in the case of the Noetherian condition, in countable algebraic structures it is possible to prove the existence of maximal elements without AC. This raises the question: if we do not need AC to prove a certain theorem, exactly what assumption about infinity do we need? In recent decades, **reverse mathematics** has found very precise answers to this question for many classical theorems of analysis, topology, and algebra. It has also found, remarkably, that the *same* few assumptions about infinity turn out to be the "right axioms" to prove most of the well-known theorems.

In the case of algebra, reverse mathematics is concerned with *countable* structures, particularly countable rings, fields, and groups. The foundation of any study of countable structures is the theory of the natural numbers; that is, arithmetic. However, we now have to describe arithmetic very precisely, in a language that typically has the following ingredients:

1. The constant 0 and the function symbol $S$ (for successor), which give us names $0, S0, SS0, \ldots$ for the natural numbers.
2. The symbols $+$ and $\cdot$ for sum and product.
3. Variables $a, b, c, \ldots, x, y, z, \ldots$ for natural numbers and $X, Y, Z, \ldots$ for sets of natural numbers.
4. Logic symbols: $\wedge, \vee, \neg, \Rightarrow$ for "and," "or," "not," "implies," the quantifiers $\forall$ and $\exists$ for "for all" and "there exists," and the parentheses ( and ).
5. The symbols $\in$ for membership and $=$ for equality.

Given some rules for writing meaningful formulas, quite obvious but tedious to enumerate, we can then write down the **Peano axioms** for arithmetic:

$$\forall x(0 \neq Sx) \qquad\qquad\qquad\qquad\qquad \text{(0 is not a successor)}$$

$$\forall x \forall y(x \neq y \Rightarrow Sx \neq Sy) \qquad\qquad\qquad \text{(S is injective)}$$

$$\forall x \forall y(x + 0 = x \wedge x + Sy = S(x + y)) \qquad \text{(inductive definition of } +)$$

$$\forall x \forall y(x \cdot 0 = 0 \wedge x \cdot Sy = x \cdot y + x) \qquad \text{(inductive definition of } \cdot)$$

$$\forall X \Big( \big(0 \in X \wedge \forall n(n \in X \Rightarrow n + 1 \in X)\big) \Rightarrow \forall n(n \in X) \Big)$$
$$\text{(induction axiom)}$$

The induction axiom axiom formalizes the "base step, induction step" idea of induction. It says that if X includes 0, and if X includes $n + 1$ when it includes $n$, then $X$ includes all natural numbers.

To these we need to add a **set existence axiom**, because as yet we have no axiom allowing us to prove that sets of natural numbers exist. A natural axiom to choose is one saying that a set exists if it has a *definition* in the language of arithmetic. For example, the following formula $\varphi(x)$ defines "x is an even number":

$$\exists y(x = SS0 \cdot y).$$

We might also want to talk about the even numbers in a given set $Z$. These are defined by the formula

$$x \in Z \wedge \exists y(x = SS0 \cdot y).$$

Motivated by these examples, we state the **arithmetic comprehension axiom**:

$$\exists X(n \in X \Leftrightarrow \varphi(n)),$$

where $\varphi(n)$ is any formula in the language of arithmetic with no quantified set variables and not including the set variable $X$. We exclude $X$ from $\varphi$ to avoid circularity – defining $X$ in terms of itself – and we do not allow the other set variables $Z$ in $\varphi$ to be quantified, since that too is circular. (If $X$ depends on a condition involving all sets, then $X$ depends on itself.)

In this axiom system it is possible to prove the countable versions of the maximal ideal and vector space basis theorems mentioned in the previous subsection:

- Any nonzero countable ring has a maximal ideal.
- Any countable vector space has a basis.

Moreover, each of these theorems *implies* the arithmetic comprehension axiom, so the arithmetic comprehension axiom is the "right axiom" to prove them. This was proved by Friedman et al. (1983), and the proofs may be seen in Simpson (2009).

The system of Peano axioms plus arithmetic comprehension is called $ACA_0$. There are also two noteworthy weaker systems, $RCA_0$ and $WKL_0$, which are more complicated to describe. In the weakest system, $RCA_0$, only "computable" sets are assumed to exist. Rather surprisingly, $RCA_0$ is strong enough to prove the fundamental theorem of algebra. If Kronecker had known this, perhaps he might have changed his mind about FTA! For an introduction to $ACA_0$, $WKL_0$, and $RCA_0$, see Stillwell (2018).

# 6

# Vector Spaces

## Preview

The difference between fields and rings is reflected by a difference in the kinds of space for which they can serve as "coordinates." A space with coordinates from a field $F$ is called a **vector space** ("over" $F$, or an $F$-vector space) and a space with coordinates from a ring $R$ is called a **module** (an $R$-module). Of course there are similarities between vector spaces and modules, but vector spaces are easier to work with. So we discuss vector spaces in this chapter, and in Chapter 8 we see how far their properties carry over to modules.

The key properties of a vector space are the existence of a **basis** and an associated **dimension**. These properties behave as expected with respect to **subspaces**: if $U$ is a subspace of a vector space $V$, then $U$ has a basis no larger than any basis of $V$. Dimension also behaves properly with respect to **linear maps**, the maps that preserve vector sums and scalar multiples, and which may be represented by **matrices**.

When vector spaces are used in number theory, it is common to vary the field $F$ of scalars, and to view one field $E$ *relative to*, or "over" a subfield $F$. This leads to viewing $E$ as a vector space over $F$. A basis of this vector space is then called a "basis for $E$ over $F$," its dimension is the "dimension of $E$ over $F$," and so on. Typically, $E = \mathbb{Q}(\alpha)$ for some algebraic number $\alpha$. Then to investigate some $\beta \in E$, we may consider $F = \mathbb{Q}(\beta)$ and look at both $E$ over $F$ and $F$ over $\mathbb{Q}$, and the associated bases and dimensions.

Since this approach to linear algebra is more general than the typical undergraduate course, we spend some time reviewing basic facts such as existence of bases in this chapter, and continue in the next with the deeper theory surrounding the **determinant**.

## 6.1 Vector Space Basis and Dimension

I expect that most readers of this book have had a course in linear algebra, with linear equations, matrices, linear maps, determinants, and perhaps a concept of vector space. Very likely, the coordinates of all the spaces, matrix entries, and coefficients of equations were real numbers, because these are the numbers relevant to most users of linear algebra. However, solving linear equations requires only the operations of addition, subtraction, multiplication, and division, so linear algebra can take place over any field $F$. Most of what the reader already knows can be applied to any field. The basic definitions are:

**Definitions.** A **vector space** over a field $F$, or an $F$-**vector space**, is a set $V$ of objects called **vectors** which can be added and multiplied by members of $F$. (Members of $F$ are also called **scalars** in this context.) $V$ includes a **zero vector 0** and, for each vector $v$, an **additive inverse** $-v$.

Vector addition has the properties of an **Abelian group** (written in additive notation, as in Section 3.7), namely

$$u + v = v + u,$$
$$u + (v + w) = (u + v) + w,$$
$$u + 0 = u,$$
$$u + (-u) = 0.$$

Scalar multiplication has the properties, for any $a, b \in F$,

$$a(bu) = (ab)u,$$
$$1u = u,$$
$$a(u + v) = au + av,$$
$$(a + b)u = au + bu.$$

Thus, although addition and multiplication now have a more general meaning, their properties are like those for ordinary addition and multiplication.[1] This makes it easy to compute with vectors and scalars, as long as one does not generally attempt to multiply or divide vectors!

---

[1] In fact, when the field $\mathbb{R}$ of real numbers is viewed as a vector space over *itself*, addition and scalar multiplication are just the ordinary addition and multiplication, because the real numbers are both vectors and scalars.

In abstract algebra, the option to choose the field $F$ is important, because more than one field may be under consideration at the same time. For example, the field

$$\mathbb{Q}(\sqrt{2}) = \{a + b\sqrt{2} : a, b \in \mathbb{Q}\}$$

can be viewed as a vector space over itself, by viewing members of $\mathbb{Q}(\sqrt{2})$ as both "vectors" and "scalars." However, it is more enlightening to view $\mathbb{Q}(\sqrt{2})$ as a vector space over the field $\mathbb{Q}$, so the "scalars" are the rational numbers. From this viewpoint, it seems that $\mathbb{Q}(\sqrt{2})$ is **two-dimensional** relative to $\mathbb{Q}$, because each $a + b\sqrt{2} \in \mathbb{Q}(\sqrt{2})$ has two rational "coordinates," $a$ and $b$.

This concept of dimension is formalized via the following definitions, which extend those given in Section 4.5 in the special case where fields are viewed as vector spaces.

**Definitions.** Members $v_1, \ldots, v_n$ of a vector space $V$ over a field $F$ are said to be **linearly independent** if

$$a_1 v_1 + \cdots + a_n v_n = 0, \quad \text{where } a_1, \ldots, a_n \in F,$$

only if $a_1 = \cdots = a_n = 0$.

Members $v_1, \ldots, v_n$ are said to **span** $V$ if, for each $v \in V$,

$$v = a_1 v_1 + \cdots + a_n v_n \quad \text{for some } a_1, \ldots, a_n \in F.$$

A **finite basis** for $V$ over $F$ is a linearly independent set of vectors $v_1, \ldots, v_n$ in $V$ that span $V$. In this case $V$ is said to be **finite-dimensional**, and the **dimension** of $V$ over $F$ is $n$.

According to these definitions, the vector space $\mathbb{Q}(\sqrt{2})$ has basis with members 1 and $\sqrt{2}$, so it has dimension 2 over $\mathbb{Q}$, as anticipated. (We pointed out in Section 4.5 that 1 and $\sqrt{2}$ are linearly independent over $\mathbb{Q}$, because $\sqrt{2}$ is irrational.) More generally, an algebraic number field $\mathbb{Q}(\alpha)$ of degree $n$ has dimension $n$ as a vector space over $\mathbb{Q}$ because it has basis $1, \alpha, \ldots, \alpha^{n-1}$, as we showed in Section 4.5.

The existence of a basis for an arbitrary vector space over an arbitrary field is more problematic, because it is equivalent to the axiom of choice. A famous example is $\mathbb{R}$ as a vector space over $\mathbb{Q}$: even here, existence of a basis depends on the axiom of choice. Fortunately, the bases we want to study in this book are finite. However, we still need to check that "dimension" is well defined, because it is conceivable that finite bases of different sizes exist.

## Exercises

The idea of viewing $\mathbb{R}$ as a vector space over $\mathbb{Q}$ goes back to Hamel (1905). He used it to create a discontinuous function $f$ such that $f(x + y) = f(x) + f(y)$ for any $x, y \in \mathbb{R}$, a property previously known to hold only for the continuous functions $f(x) = kx$.

A **Hamel basis** for $\mathbb{R}$ over $\mathbb{Q}$ is obtained by a "maximal" argument like the one used in the exercises to Section 5.4 to prove the existence of maximal ideals.

1. Suppose that $B$ is a set of real numbers that are linearly independent over $\mathbb{Q}$. That is, if $x_1, \ldots, x_n \in B$ and $q_1, \ldots, q_n \in \mathbb{Q}$, then

$$q_1 x_1 + \cdots + q_n x_n = 0 \implies q_1 = \cdots = q_n = 0.$$

   Show that if $B$ is not a basis, then there is an independent set $B' \supseteq B$ whose span includes some real $x'$ not in the span of $B$.
2. Suppose $\mathcal{B}$ is a collection of independent sets $B_i$ that is linearly ordered by $\subseteq$; that is, if $B_i, B_j \in \mathcal{B}$, then either $B_i \subseteq B_j$ or $B_j \subseteq B_i$. Show that if $x_1, \ldots, x_n$ are members of sets in $\mathcal{B}$, then in fact $x_1, \ldots, x_n$ belong to a single $B_i \in \mathcal{B}$.
3. Deduce that the union $B$ of all sets $B_i \in \mathcal{B}$ is also an independent set, so $B$ is an upper bound to $\mathcal{B}$.

   Now according to Zorn's lemma (see Section 5.4), the collection $\mathcal{C}$ of all independent sets of real numbers has a maximal element, $C$.

4. Deduce from exercise 1 that $C$ is a basis for $\mathbb{R}$ over $\mathbb{Q}$.
5. By the same argument, show that any field $E$ has a basis over a subfield $F \subseteq E$.

   We can now use the Hamel basis $B$ of $\mathbb{R}$ over $\mathbb{Q}$, as Hamel did, to find a discontinuous function $f$ such that $f(x + y) = f(x) + f(y)$.

6. Let $x_0$ be any particular member of $B$, and let $f(x)$ be the coefficient of $x_0$ in the expression for $x$ in its unique expression $q_1 x_1 + \cdots + q_n x_n = 0$ as a linear combination of members of $B$ with rational coefficients (taking $f(x) = 0$ if $x_0$ does not occur in this expression). Explain why $f(x + y) = f(x) + f(y)$.
7. Also explain why $f$ is not continuous.

## 6.2  Finite-Dimensional Vector Spaces

In Section 4.5 we already appealed to the concept of dimension, when claiming that a basis of $n$ elements implies that any set of $n + 1$ elements is linearly dependent. The context was a field $E$, viewed as a vector space over a field $F$. The proof for any finite-dimensional vector space $V$ over $F$ can be done via the following lemma, often called the "Steinitz exchange lemma" but known earlier to Grassmann (1862).

**Grassmann exchange lemma.**  *If $n$ vectors span $V$, then no $n + 1$ vectors are independent.*

*Proof.* Suppose (for the sake of contradiction) that $u_1, \ldots, u_n$ span $V$ and that $v_1, \ldots, v_n, v_{n+1} \in V$ are linearly independent. The plan is to replace some $u_i$ by $v_1$ while ensuring that $v_1$ and the remaining $u_i$ still span $V$. We can likewise replace another $u_i$ by $v_2$, and so on. In the end, all the $u_i$ are replaced, and we have a spanning set $v_1, \ldots, v_n$. In particular, $v_{n+1}$ is a linear combination of $v_1, \ldots, v_n$, contrary to the assumption of linear independence.

To see why this plan succeeds, suppose we have replaced $m - 1$ of the $u_i$ by $v_1, \ldots, v_{m-1}$, so now we want to replace another $u_i$ by $v_m$. Success so far means that $v_1, \ldots, v_{m-1}$ and the remaining $u_i$ span $V$, so

$$v_m = a_1 v_1 + \cdots + a_{m-1} v_{m-1} + \text{terms } b_i u_i$$

where $a_1, \ldots, a_{m-1}$ and the $b_i$ are in $F$. Since $v_m$ is not a linear combination of $v_1, \ldots, v_{m-1}$ (by their linear independence), some $b_i \neq 0$. But then, dividing by $b_i$, we get $u_i$ as a linear combination of $v_1, \ldots, v_m$ and the remaining $u_j$. So $u_i$ can be replaced by $v_m$, and we still have a spanning set.

Thus, the exchange process can be continued until all the $u_i$ are replaced, leading to the contradiction foreshadowed above. This proves the exchange lemma.                                                                                    □

**Corollary.**  *If $V$ has a basis over $F$ with $n$ elements, then all bases of $V$ over $F$ have $n$ elements. (Vector space dimension is well defined.)*

*Proof.* Suppose that $u_1, \ldots, u_n$ is a basis for $V$ over $F$. Then any larger set is not linearly independent, by the lemma, and hence not a basis. Similarly, no smaller set is a basis either, otherwise $u_1, \ldots, u_n$ would not be a basis, by the lemma again.                                                                                    □

Finite-dimensionality has important consequences when $V$ is an extension field $E \supseteq F$. In this case, as we saw in Section 4.5, if $E$ has dimension $n$, then each $x \in E$ is algebraic of degree $\leq n$ over $F$. This is because the $n + 1$

elements $1, x, x^2, \ldots, x^n$ are linearly dependent over $F$, which means that $x$ satisfies an equation of degree $n$ with coefficients in $F$.

Moreover, if $D \supseteq E \supseteq F$ are fields, with each extension of finite dimension, then we can calculate the dimension of $D$ over $F$ from the dimensions of $D$ over $E$ and of $E$ over $F$. This was noted by Dedekind (1894) without proof. The proof in fact is very simple, underlining the power of the basis concept. Since the key concept of relative dimension will be useful later, we introduce the following:

**Notation.** If $F \supseteq E$ is a field extension of finite dimension, we denote the dimension of $E$ over $F$ by $[E : F]$.

If one does not think in terms of bases, then it is hard to see that the number $\sqrt[3]{2} + \sqrt[5]{7}$, say, is even algebraic. The number $\sqrt[3]{2}$ belongs to a field of degree 3, hence of dimension 3, over $\mathbb{Q}$, and $\sqrt[5]{7}$ similarly belongs to a field of dimension 5 over $\mathbb{Q}$, but what kind of a field contains their sum? The answer is a field of degree 15, by the following theorem.

**Dedekind product theorem.** *If $D \supseteq E \supseteq F$ are fields, such that $[D : E] = m$ and $[E : F] = n$, then $[D : F] = mn = [D : E][E : F]$.*

*Proof.* Let $\delta_1, \ldots, \delta_m$ be a basis for $D$ over $E$, so any $d \in D$ can be written

$$d = e_1 \delta_1 + \cdots + e_m \delta_m \quad \text{for some } e_1, \ldots, e_m \in E \qquad (*)$$

and let $\epsilon_1, \ldots, \epsilon_n$ a basis for $E$ over $F$, so any $e \in E$ can be written

$$e = f_1 \epsilon_1 + \cdots + f_n \epsilon_n \quad \text{for some } f_1, \ldots, f_n \in F. \qquad (**)$$

It follows, since each $e_i$ in $(*)$ can be rewritten in terms of the basis elements $\epsilon_j$ as in $(**)$, that the $mn$ elements $\delta_i \epsilon_j$ span $D$ over $F$.

Also, the elements $\delta_i \epsilon_j$ of $D$ are linearly independent over $F$. If a sum of terms $f_{ij} \delta_i \epsilon_j$ equals 0 with the $f_{ij} \in F$, then we have

$$
\begin{aligned}
0 = \ & (f_{11} \epsilon_1 + \cdots + f_{1n} \epsilon_n) \delta_1 \\
& + (f_{21} \epsilon_1 + \cdots + f_{2n} \epsilon_n) \delta_2 \\
& \vdots \\
& + (f_{m1} \epsilon_1 + \cdots + f_{mn} \epsilon_n) \delta_m.
\end{aligned}
$$

Since the $\delta_i$ are linearly independent over $E$, their coefficients

$$f_{i1} \epsilon_1 + \cdots + f_{in} \epsilon_n = 0.$$

Then, since the $\epsilon_j$ are also linear independent over $F$, their coefficients $f_{ij} = 0$. This shows that the $mn$ elements $\delta_i \epsilon_j$ are linearly independent, and hence they form a basis for $D$ over $F$.                                         □

It follows that the sum of any two algebraic numbers, $\alpha$ of degree $m$ and $\beta$ of degree $n$, respectively, is an algebraic number of degree $\leq mn$. This is because $\alpha$ belongs to a field $E$ of dimension $m$ over $\mathbb{Q}$, in which case $\alpha + \beta$ belongs to the field $D = E(\beta)$, which is of degree $\leq n$ over $E$, and hence of degree $\leq mn$ over $\mathbb{Q}$. (By Section 4.5, any member of a field of degree $d$ over $\mathbb{Q}$ has degree $\leq d$.) By a similar argument, $\alpha\beta$ also has degree $\leq mn$ over $\mathbb{Q}$.

Thus, it is quite easy to prove that the sum or product of algebraic *numbers* is an algebraic number. Proving that the sum of algebraic *integers* is an algebraic integer, as in Section 4.6, was not so simple, because it invoked the determinant concept. We will need this concept, and other more sophisticated parts of linear algebra, in the study of algebraic integers and their generalizations.

### 6.2.1 Subspaces

**Definition.** A **subspace** $U$ of an $F$-vector space $V$ is a nonempty subset of $V$ closed under vector addition and multiplication by members of $F$.

It is easy to see that a subspace $U$ of $V$ is itself an $F$-vector space. We naturally expect that dimension of $U \leq$ dimension of $V$, and this is correct. Indeed, it is a corollary of the Grassmann exchange lemma.

**Dimension of a subspace.** *If $U$ is a subspace of a finite-dimensional vector space $V$, then dimension of $U \leq$ dimension of $V$.*

*Proof.* If $V$ has dimension $n$, then any $n + 1$ vectors in $U$ are also in $V$, and hence linearly dependent by the Grassmann exchange lemma. Skipping the trivial case $U = \{\mathbf{0}\}$, we find a basis of $U$ by the following process.

Choose any nonzero $\boldsymbol{u}_1 \in U$. If $\boldsymbol{u}_1$ does not span $U$, choose a $\boldsymbol{u}_2 \in U$ not in the span of $\boldsymbol{u}_1$. If $\boldsymbol{u}_1, \boldsymbol{u}_2$ do not span $U$, choose $\boldsymbol{u}_3 \in U$ not in the span of $\boldsymbol{u}_1, \boldsymbol{u}_2$, and so on. Then we have an independent set at each stage, because if $\boldsymbol{u}_1, \ldots, \boldsymbol{u}_k$ are independent and $\boldsymbol{u}_1, \ldots, \boldsymbol{u}_k, \boldsymbol{u}_{k+1}$ are not, $\boldsymbol{u}_{k+1}$ has a nonzero coefficient $a_{k+1} \in F$ in a relation

$$a_1 \boldsymbol{u}_1 + \cdots + a_k \boldsymbol{u}_k + a_{k+1} \boldsymbol{u}_{k+1} = 0, \quad \text{where } a_1, \ldots, a_k, a_{k+1} \in F,$$

which means $\boldsymbol{u}_{k+1} = -a_{k+1}^{-1}(a_1 \boldsymbol{u}_1 + \cdots + a_k \boldsymbol{u}_k)$ is in the span of $\boldsymbol{u}_1, \ldots, \boldsymbol{u}_k$. Thus, the process must terminate before we get $n + 1$ vectors.

At termination we have an independent spanning set with $< n + 1$ vectors, so $U$ has dimension $\leq n$.                                              □

This proof depends on $F$ being a field, because we divide by $a_{k+1}$ in the final step. The corresponding statement with a ring $R$ in place of $F$ does *not* hold for all rings $R$, though it holds for certain rings such as $\mathbb{Z}$, as we will see in Section 8.4. We will also see there that the proof can avoid division, but at the cost of a certain amount of gymnastics with addition, subtraction, and multiplication.

## Exercises

We can now identify *dimension* with *degree*, as foreshadowed in Section 4.5, where the degree of an algebraic number $\alpha$ is the degree of the irreducible polynomial (which is unique up to constant factors) satisfied by $\alpha$.

A famous application of dimension, and the Dedekind product theorem in particular, concerns the numbers $\alpha$ obtained from rational numbers by *nested square roots*.

1.  Observing that $\alpha = \sqrt{1 + \sqrt{2}}$ satisfies the equation
    $$\alpha^2 - \left(1 + \sqrt{2}\right) = 0,$$
    with coefficients in $\mathbb{Q}(\sqrt{2})$, explain why the field $\mathbb{Q}(\alpha)$ is of degree either 1 or 2 over $\mathbb{Q}(\sqrt{2})$.[2]
2.  Deduce that $\mathbb{Q}(\alpha)$ is of degree 2 or 4 over $\mathbb{Q}$.
3.  Generalizing the argument in the previous two questions, explain why a number $\beta$ obtained by nested square roots from rationals generates a field $\mathbb{Q}(\beta)$ of degree $2^m$ over $\mathbb{Q}$.
4.  Show, from the irrationality of $\sqrt[3]{2}$, that $\mathbb{Q}(\sqrt[3]{2})$ is of degree 3 over $\mathbb{Q}$, and hence that $\sqrt[3]{2}$ does not belong to any field obtained by square roots.

The latter result implies that $\sqrt[3]{2}$ is not obtainable from rational numbers by the rational operations $+, -, \times, \div$ and square roots. This answers an ancient geometric question called *duplication of the cube*, by showing that the length $\sqrt[3]{2}$ cannot be obtained from the unit length by a so-called *straightedge and compass construction*. (For more information on the algebra of geometric constructions, see for example Stillwell (1994).)

---

[2] It may not be clear whether $\alpha$ is in $\mathbb{Q}(\sqrt{2})$ or not, but it doesn't matter. Its degree over $\mathbb{Q}(\sqrt{2})$ is *at most* 2.

## 6.3 Linear Maps

Linear maps are maps that preserve "vector space structure" in the sense of the following definitions.

**Definitions.** If $V$ and $V'$ are $F$-vector spaces, then a map $f : V \to V'$ is called **linear** (or, if we want to emphasize the field of scalars, $F$-*linear*) if it preserves vector sums and scalar multiples, that is:

$$f(v_1 + v_2) = f(v_1) + f(v_2) \quad \text{and} \quad f(av) = af(v)$$
$$\text{for any } v, v_1, v_2 \in V, \ a \in F.$$

A linear map $f : V \to V'$ is also called a vector space **homomorphism**. If $f$ is also **injective** (one-to-one) and **surjective** (onto), then it is called an **isomorphism** or, specifically, a **vector space isomorphism**.

A homomorphism (which comes from the Greek for something like "similar form") also preserves the zero vector and additive inverses. For example

$$v + 0 = v \Rightarrow f(v + 0) = f(v)$$
$$\Rightarrow f(v) + f(0) = f(v), \qquad \qquad \text{by linearity}$$
$$\Rightarrow f(0) = 0. \qquad \text{by uniqueness of zero vector}$$

An isomorphism (from the Greek for "same form") preserves not only sums and scalar multiples, but also *dimension*, as the following theorem shows.

**Invariance of dimension under isomorphisms.** *If $V$ has a finite basis, a vector space isomorphism $f : V \to V'$ maps a basis of $V$ to a basis of $V'$, necessarily of the same size.*

*Proof.* If $v_1, \ldots, v_n$ is a basis of $V$, then $f(v_1), \ldots, f(v_n)$ is a basis of $V$ because:

1. Since $v_1, \ldots, v_n$ span $V$, we can write any $v \in V$ in the form

$$v = a_1 v_1 + \cdots + a_n v_n \quad \text{for some } a_1, \ldots, a_n \in F.$$

It follows by linearity that

$$f(v) = a_1 f(v_1) + \cdots + a_n f(v_n).$$

The latter equation shows that $f(v_1), \ldots, f(v_n)$ span $V'$, because $f$ is onto $V'$ and hence any member of $V'$ is of the form $f(v)$ for some $v \in V$.

2. To see why $f(v_1), \ldots, f(v_n)$ are linearly independent, notice that

$$\mathbf{0} = a_1 f(v_1) + \cdots + a_n f(v_n) \qquad \text{for some } a_1, \ldots, a_n \in F$$
$$\Rightarrow \mathbf{0} = f(a_1 v_1 + \cdots + a_n v_n) \qquad \text{by linearity}$$
$$\Rightarrow f(\mathbf{0}) = f(a_1 v_1 + \cdots + a_n v_n) \qquad \text{since } \mathbf{0} = f(\mathbf{0})$$
$$\Rightarrow \mathbf{0} = a_1 v_1 + \cdots + a_n v_n \qquad \text{since } f \text{ is injective}$$
$$\Rightarrow a_1 = \cdots = a_n = 0 \qquad \text{since } v_1, \ldots, v_n \text{ are linearly independent.}$$

Thus $f(v_1), \ldots, f(v_n)$ is a basis of $V'$, so both $V$ and $V'$ have dimension $n$.

$\square$

**Corollary.** *If $V$ has a finite basis, a linear map $f : V \to V$ is injective if and only if it is surjective.*

*Proof.* If $f$ is injective, then part 2 of the proof above shows $f(v_1), \ldots, f(v_n)$ are linearly independent. If they do *not* span, then some $v \in V$ is not in their span, in which case $v, f(v_1), \ldots, f(v_n)$ are independent, contrary to the dimension of $V$ being $n$. So in fact, for any $v \in V$,

$$v = a_1 f(v_1) + \cdots + a_n f(v_n) = f(a_1 v_1 + \cdots + a_n v_n)$$

for some $a_1, \ldots, a_n \in F$. That is, $f$ is surjective.

If $f$ is surjective, then part 1 of the proof above shows that $f(v_1), \ldots, f(v_n)$ span $V$. If $f$ is *not* injective, then

$$f(a_1 v_1 + \cdots + a_n v_n) = a_1 f(v_1) + \cdots + a_n f(v_n) = \mathbf{0},$$

for some $a_1, \ldots, a_n \in F$, not all zero, in which case some $f(v_i)$ is a linear combination of the other $f(v_j)$. So the other $f(v_j)$ span $V$, again contrary to the dimension of $V$ being $n$.

$\square$

### 6.3.1 Matrices

It follows from the calculations in the theorem on invariance of dimension that the value of a linear map $f$ on any vector $v$ is determined by its values $f(v_1), \ldots, f(v_n)$ on a set of basis vectors $v_1, \ldots, v_n$. Thus, if $a_{ij}$ are the elements of $F$ such that

$$f(v_1) = a_{11} v_1 + \cdots + a_{1n} v_n$$
$$\vdots$$
$$f(v_n) = a_{n1} v_1 + \cdots + a_{nn} v_n$$

then $f$ is determined by the **matrix**

$$A = \begin{pmatrix} a_{11} & \cdots & a_{1n} \\ \vdots & & \vdots \\ a_{n1} & \cdots & a_{nn} \end{pmatrix}.$$

We will assume that the basic properties of matrices are known from an introductory linear algebra course, starting with the fact that matrix multiplication reflects composition of linear maps. It is important to remember, however, that the matrix of a linear map depends on a choice of basis. Sometimes we wish to use properties of a matrix that are *independent of basis*, and hence intrinsic properties of the corresponding linear map.

Two basis-independent quantities are the **trace** and the **determinant** of a matrix $A$, written as $\mathrm{Tr}(A)$ and $\det(A)$. Both of these invariants involve only the ring operations of sum, difference, and product, as we will see in Section 7.5. Therefore, they also apply in a setting where we have only a ring in place of a field. This is the setting of *modules*, which we study in Chapter 8.

### Exercises

The basis-independence of the determinant follows from basic properties of matrices and determinants, such as $\det(AB) = \det(A)\det(B)$. They are probably familiar from an introductory linear algebra course, so we will assume them for this chapter. However, we will discuss determinants more fully in the next chapter (and revisit these exercises in Section 7.4), because determinants are really too remarkable to be taken for granted.

1.  Show that the change of basis effected by a matrix $B$ causes the transformation expressed by matrix $A$ to be expressed by matrix $BAB^{-1}$.
2.  Show also that $\det(B)^{-1} = \det(B^{-1})$.
3.  Hence, conclude that $\det(BAB^{-1}) = \det(A)$.

## 6.4 Algebraic Numbers as Matrices

In Section 4.4 we saw how algebraic numbers are represented "concretely" – in terms of rational numbers – as congruence classes of polynomials with rational coefficients. Another way to realize algebraic numbers "concretely," in this sense, is by means of matrices. We mention this here because it involves an interesting example of a linear map, which we generalize in Section 7.5.

A simple example of a matrix that behaves like an algebraic number is

$$i = \begin{pmatrix} 0 & -1 \\ 1 & 0 \end{pmatrix}$$

Multiplying $i$ by itself we get

$$i^2 = \begin{pmatrix} 0 & -1 \\ 1 & 0 \end{pmatrix} \begin{pmatrix} 0 & -1 \\ 1 & 0 \end{pmatrix} = \begin{pmatrix} -1 & 0 \\ 0 & -1 \end{pmatrix} = -\mathbf{1},$$

where $\mathbf{1}$ denotes the $2 \times 2$ identity matrix. Thus, in the world of $2 \times 2$ matrices, $i$ is a solution of the equation $x^2 = -1$, or $x^2 + 1 = 0$.

Now suppose we have an algebraic number $\alpha$, with minimal equation

$$x^n + a_{n-1}x^{n-1} + \cdots + a_1x + a_0 = 0, \quad \text{where } a_0, \ldots, a_{n-1} \in \mathbb{Q}. \quad (*)$$

Part 3 of the proof in Section 4.4 shows that $\mathbb{Q}(\alpha)$ equals the polynomial ring $\mathbb{Q}[\alpha]$, which in turn consists of all linear combinations of $1, \alpha, \alpha^2, \ldots, \alpha^{n-1}$ with coefficients in $\mathbb{Q}$, since any polynomial is congruent to one of degree less than $n$, mod $p(x)$. For example $\alpha^n = -a_{n-1}\alpha^{n-1} - \cdots - a_1\alpha - a_0$.

Now the operation $m_\alpha$ of **multiplication by** $\alpha$ is a $\mathbb{Q}$-linear map because

$$m_\alpha(ru + sv) = rm_\alpha(u) + sm_\alpha(v) \quad \text{for any } r, s \in \mathbb{Q} \text{ and } u, v \in \mathbb{Q}(\alpha).$$

Thus, $m_\alpha$ is determined by what it does to the $n$-tuple $1, \alpha, \alpha^2, \ldots, \alpha^{n-1}$. Indeed, when $n$-tuples are written as column vectors, $m_\alpha$ sends

$$\begin{pmatrix} 1 \\ \alpha \\ \alpha^2 \\ \vdots \\ \alpha^{n-1} \end{pmatrix} \quad \text{to} \quad \begin{pmatrix} \alpha \\ \alpha^2 \\ \alpha^3 \\ \vdots \\ -a_{n-1}\alpha^{n-1} - \cdots - a_1\alpha - a_0 \end{pmatrix}.$$

This allows us to view multiplication by $\alpha$ as the matrix

$$\alpha = \begin{pmatrix} 0 & 1 & 0 & \cdots & 0 & 0 \\ 0 & 0 & 1 & \cdots & 0 & 0 \\ \vdots & \vdots & \vdots & & \vdots & \vdots \\ 0 & 0 & 0 & \cdots & 0 & 1 \\ -a_0 & -a_1 & -a_2 & \cdots & -a_{n-2} & -a_{n-1} \end{pmatrix},$$

because left multiplication of the $n$-tuple $(1, \alpha, \alpha^2, \ldots, \alpha^{n-1})$ by $\alpha$ yields the required $n$-tuple $(\alpha, \alpha^2, \alpha^3, \ldots, -a_{n-1}\alpha^{n-1} - \cdots - a_1\alpha - a_0)$.

It follows, since composition of $m_\alpha$ with itself corresponds to multiplication by powers of $\alpha$, and scalar multiples and sums of powers correspond to matrix scalar multiples and sums, that matrix polynomials in $\alpha$ behave exactly the same as polynomials in $\alpha$. In particular,

$$\alpha^n + a_{n-1}\alpha^{n-1} + \cdots + a_1\alpha + a_0 \mathbf{1} = \mathbf{0} \quad \text{(the zero matrix).}$$

For example, the (transpose of the) matrix

$$x = \begin{pmatrix} 0 & 2 \\ 1 & 0 \end{pmatrix},$$

for which $a_0 = -2$, ought to satisfy the equation $x^2 - 2 \cdot \mathbf{1} = \mathbf{0}$. And indeed,

$$x^2 = \begin{pmatrix} 0 & 2 \\ 1 & 0 \end{pmatrix}\begin{pmatrix} 0 & 2 \\ 1 & 0 \end{pmatrix} = \begin{pmatrix} 2 & 0 \\ 0 & 2 \end{pmatrix} = 2 \cdot \mathbf{1}.$$

Notice that the matrix $\alpha$ has rational entries if $\alpha$ is an algebraic number, and indeed, integer entries if $\alpha$ is an algebraic integer, because the minimal equation (*) for $\alpha$ has integer coefficients in the latter case, by the corollary in Section 4.7. In particular, we can represent the number $\sqrt{2}$ by the integer matrix $\begin{pmatrix} 0 & 2 \\ 1 & 0 \end{pmatrix}$.

## Exercises

Representation of algebraic numbers by matrices allows us to approach their norms via the concept of determinant. From this point of view, the multiplicative property of norm follows from the multiplicative property of determinant: $\det(AB) = \det(A)\det(B)$.

1. Show that the complex number $a + bi$ (where $a, b \in \mathbb{Q}$ or $\mathbb{R}$) is represented by the matrix

$$\begin{pmatrix} a & -b \\ b & a \end{pmatrix}, \text{ whose determinant equals } a^2 + b^2 = |a + bi|^2.$$

2. Show that the element $a + b\sqrt{2}$ of $\mathbb{Q}(\sqrt{2})$ is represented by the matrix, with $a, b \in \mathbb{Q}$,

$$\begin{pmatrix} a & 2b \\ b & a \end{pmatrix}, \text{ whose determinant equals } a^2 - 2b^2.$$

3. Conclude that the norm $a^2 - 2b^2$ in $\mathbb{Q}(\sqrt{2})$ is multiplicative.

4. Using the matrix for multiplication by $2^{1/3}$ in $\mathbb{Q}(2^{1/3})$, show that the element $a + b \cdot 2^{1/3} + c \cdot 2^{2/3}$ of $\mathbb{Q}(2^{1/3})$ is represented by the matrix

$$\begin{pmatrix} a & 2c & 2b \\ b & a & 2c \\ c & b & a \end{pmatrix}.$$

5. Hence, show that the norm of $a + b \cdot 2^{1/3} + c \cdot 2^{2/3}$ is $a^3 + 2b^3 + 4c^3 - 6abc$.

The multiplicative property of det is remarkably powerful, as can be seen in the case of the **quaternion** matrices

$$Q(a,b,c,d) = \begin{pmatrix} a + bi & c + di \\ -c + di & a - bi \end{pmatrix},$$

whose determinant equals $a^2 + b^2 + c^2 + d^2$.

6. Writing each quaternion matrix in the form

$$\begin{pmatrix} \alpha & \beta \\ -\bar{\beta} & \bar{\alpha} \end{pmatrix}, \quad \text{where } \alpha = a + bi \text{ and } \beta = c + di,$$

check that the product of quaternion matrices is a quaternion matrix.

7. Deduce, because $\det(Q_1)\det(Q_2) = \det(Q_1 Q_2)$, that there is a *four square identity*, expressing the product $(a_1^2 + b_1^2 + c_1^2 + d_1^2)$ $(a_2^2 + b_2^2 + c_2^2 + d_2^2)$ as a sum of four squares. Such an identity was first discovered by Euler (1748).

## 6.5 The Theorem of the Primitive Element

If $\alpha$ is an algebraic number of degree $n$, we know that the field $\mathbb{Q}(\alpha)$ is of dimension $n$ over $\mathbb{Q}$, with basis $1, \alpha, \ldots, \alpha^{n-1}$. There is a converse to this result, called the **theorem of the primitive element**. It states that if $E \supseteq \mathbb{Q}$ is an extension of dimension $n$, then $E = \mathbb{Q}(\alpha)$ for some algebraic number $\alpha$ of degree $n$, and hence $E$ is a vector space over $\mathbb{Q}$ with basis $1, \alpha, \ldots, \alpha^{n-1}$.

To prove the theorem, it is convenient to use **derivatives** of polynomials in $F[x]$, where $F$ is an algebraic number field.

**Definition.** For any polynomial $p(x) = a_n x^n + a_{n-1} x^{n-2} \cdots + a_1 x + a_0$, we define its **derivative** to be $p'(x) = na_n x^n + (n-1)a_{n-1} x^{n-1} + \cdots + a_1$.

Thus, the derivative $p'(x)$ is the *same polynomial one finds by calculus* when $p(x)$ is in $\mathbb{R}[x]$ or $\mathbb{C}[x]$. In algebra the derivative of a polynomial is introduced by definition, because we may be working over a field $F$ where the processes of calculus do not make sense. Fortunately, we can also give algebraic proofs of the usual rules of differentiation, as long as they are restricted to polynomials. For example, we can prove the **product rule**:

For polynomials $p(x), q(x), r(x)$, if $p(x) = q(x)r(x)$, then

$$p'(x) = q'(x)r(x) + q(x)r'(x).$$

This is done by substituting general polynomials for $q(x)$ and $r(x)$, then working out both sides of the equation according to the definition.

The derivative $p'(x)$ is involved in deciding whether a polynomial $p(x)$ has a multiple root. Since $F$ is an algebraic number field, we can assume the roots of any polynomial in $F[x]$ exist in $\mathbb{C}$, by the fundamental theorem of algebra.

**Multiple root criterion.** *A polynomial $p(x) \in F[x]$ has a multiple root if and only if $\gcd(p(x), p'(x)) \neq 1$.*

*Proof.* If $\alpha$ is a root of $p(x)$, then $p(x) = (x - \alpha)q(x)$ for some polynomial $q(x)$. Now notice that $p'(x) = q(x) + (x - \alpha)q'(x)$ by the product rule. Consequently,

$\alpha$ is a multiple root of $p(x) \Leftrightarrow x - \alpha$ divides $q(x)$

$\qquad\qquad \Leftrightarrow x - \alpha$ divides $p'(x) - (x - \alpha)q'(x)$

$\qquad\qquad \Leftrightarrow x - \alpha$ divides $p'(x)$, by the product rule

$\qquad\qquad \Leftrightarrow x - \alpha$ divides $p(x)$ and $p'(x)$.

So, if $p(x)$ has a multiple root in $\mathbb{C}$, then $\gcd(p(x), p'(x)) \neq 1$, because $p(x)$ and $p'(x)$ have a common divisor $x - \alpha$ for each multiple root $\alpha$.

Conversely, if $\gcd(p(x), p'(x)) = 1$, then there is no common divisor $x - \alpha$, hence no multiple root.                                                                      □

**Corollary.** *An irreducible $p(x) \in F[x]$ has no multiple root.*

*Proof.* By the multiple root criterion, we must show $\gcd(p(x), p'(x)) = 1$ for an irreducible $p(x)$. Since $p(x)$ is irreducible, it can have a common divisor with $p'(x)$ only if $p(x)$ divides $p'(x)$. This is impossible, since $p'(x)$ has lower degree, unless $p'(x) = 0$.

But, looking at the definition of $p'(x)$, for a nonconstant polynomial $p(x)$, we see that $p'(x) \neq 0$.                                                                      □

Now, finally, we are ready to prove the theorem. The condition $E \subseteq \mathbb{C}$ enters to ensure that irreducible polynomials have distinct linear factors.

**Theorem of the primitive element.** *If $E \supseteq \mathbb{Q}$ is an extension of finite degree, and $E \subseteq \mathbb{C}$, then $E = \mathbb{Q}(\alpha)$ for some $\alpha \in E$.*

*Proof.* If $E$ is of degree $n$ over $\mathbb{Q}$, then $E = \mathbb{Q}(\alpha_1, \ldots, \alpha_n)$ where $\alpha_1, \ldots, \alpha_n$ are basis elements for $E$ over $\mathbb{Q}$. We argue by induction on $n$. If $n = 1$, then the theorem holds with $\alpha = \alpha_1$. For the induction step we can suppose $\mathbb{Q}(\alpha_1, \ldots, \alpha_{n-1}) = \mathbb{Q}(\beta)$ for some $\beta \in E$, and it remains to find $\alpha$ so that $\mathbb{Q}(\beta, \alpha_n) = \mathbb{Q}(\alpha)$.

Writing $\alpha_n$ as $\gamma$, our problem boils down to the following: given $\beta, \gamma$ that are algebraic over $\mathbb{Q}$, we wish to find $\alpha$ so that $\mathbb{Q}(\beta, \gamma) = \mathbb{Q}(\alpha)$.

We seek an element $\alpha \in \mathbb{Q}(\beta, \gamma)$ of the form $\alpha = \gamma + c\beta$ where $c \in \mathbb{Q}$. It then suffices to show that $\beta \in \mathbb{Q}(\alpha)$, because then $\gamma = \alpha - c\beta \in \mathbb{Q}(\alpha)$ too. To do this we will show that the minimal monic polynomial for $\beta$ in $\mathbb{Q}(\alpha)$ is $x - \beta$. This comes from looking at ·

$$g(x) = \text{minimal monic polynomial for } \beta \text{ over } \mathbb{Q},$$

$$h(x) = \text{minimal monic polynomial for } \gamma \text{ over } \mathbb{Q}.$$

We initially work in $\mathbb{C}$, which contains all the roots of $g(x)$ and $h(x)$. Then, since roots correspond to linear factors, over $\mathbb{C}$ we have

$$g(x) = (x - \beta_1) \cdots (x - \beta_m)$$

where $\beta_1, \ldots, \beta_m$ are the solutions of $g(x) = 0$,

$$h(x) = (x - \gamma_1) \cdots (x - \gamma_n) \quad \text{where } \gamma_1, \ldots, \gamma_n \text{ are the solutions of } h(x) = 0.$$

Now $g(x)$ has $\beta$ among its roots $\beta_1, \ldots, \beta_m$ by definition. Similarly, for $c \neq 0$, $h(\alpha - cx)$ also has the root $\beta$, because $\gamma = \alpha - c\beta$, and $\gamma$ is among the roots $\gamma_1, \ldots, \gamma_n$ of $h(x)$. This suggests that we can find the polynomial $x - \beta$ as the *greatest common divisor* of $g(x)$ and $h(\alpha - cx)$, provided we can choose $c$ so that $\beta$ is the *only* common root of $g(x)$ and $h(\alpha - cx)$.

Without loss of generality we can assume $\beta = \beta_1$ and $\gamma = \gamma_1$. Then, since $g(x)$ and $h(x)$ have no multiple roots by the corollary above, we have $\beta_i \neq \beta_1$ when $i \neq 1$ and $\gamma_j \neq \gamma_1$ when $j \neq 1$. Since $\alpha = c\beta + \gamma$, it suffices to ensure that

$$c\beta_1 + \gamma_1 \neq c\beta_i + \gamma_j \quad \text{for } i \neq 1, j \neq 1.$$

This means that $c \neq \frac{\gamma_j - \gamma_1}{\beta_1 - \beta_i}$, and we can avoid these finitely many values because $\mathbb{Q}$ is infinite.

Thus, we can indeed choose $c$ so that $\beta$ is the only common root of $g(x)$ and $h(\alpha - cx)$, which means that

$$x - \beta = \gcd(g(x), h(\alpha - cx)).$$

Now $g(x) \in \mathbb{Q}[x]$ and $h(\alpha - cx) \in \mathbb{Q}(\alpha)[x]$, because $c \in \mathbb{Q}$ and $\alpha \in \mathbb{Q}(\alpha)$. So in fact both polynomials are in $\mathbb{Q}(\alpha)[x]$, and hence the Euclidean algorithm for their gcd gives a polynomial with coefficients in $\mathbb{Q}(\alpha)$. Thus, $\beta \in \mathbb{Q}(\alpha)$, as required.                                                                                                        $\square$

## Exercises

A field whose primitive element can be found directly is $\mathbb{Q}(\sqrt{2}, \sqrt{3})$, obtained from $\mathbb{Q}(\sqrt{2})$ by adjoining the element $\sqrt{3}$. We begin by recalling from exercise 1 of Section 4.5 that $\sqrt{3}$ is not in $\mathbb{Q}(\sqrt{2})$.

1.  Deduce that $\mathbb{Q}(\sqrt{2}, \sqrt{3})$ is of dimension 2 over $\mathbb{Q}(\sqrt{2})$, and hence of dimension 4 over $\mathbb{Q}$.
2.  Show, by rational operations on $\alpha = \sqrt{2} + \sqrt{3}$ (including division), that $\sqrt{2} \in \mathbb{Q}(\sqrt{2} + \sqrt{3})$, and hence that $\sqrt{3} \in \mathbb{Q}(\sqrt{2} + \sqrt{3})$.
3.  Conclude that $\sqrt{2} + \sqrt{3}$ is a primitive element for $\mathbb{Q}(\sqrt{2}, \sqrt{3})$, and hence that $\sqrt{2} + \sqrt{3}$ is of degree 4.
4.  Show that $\sqrt{2} + \sqrt{3}$ satisfies the polynomial $x^4 - 10x^2 + 1$.
5.  Deduce that $x^4 - 10x^2 + 1$ is irreducible in $\mathbb{Q}[x]$.

## 6.6 Algebraic Number Fields and Embeddings in $\mathbb{C}$

Now that we know that each algebraic number field $E$ has the simple form $\mathbb{Q}(\alpha)$, we can relate the possible realizations of $E$ in the complex numbers $\mathbb{C}$ to the roots of the minimal polynomial $p(x)$ for $\alpha$. We saw in Section 4.4 that for each root $\alpha_i$ of $p(x)$ there is an isomorphism $\sigma_i : \mathbb{Q}(\alpha) \rightarrow \mathbb{C}$ sending $\alpha$ to $\alpha_i$. We know from Section 6.5 the roots of $p(x)$ are distinct, so if $E$ is of degree $n$, there are $n$ distinct isomorphisms $\sigma_i$.

Conversely, any isomorphism $\sigma : \mathbb{Q}(\alpha) \rightarrow \mathbb{C}$ is determined by the value of $\sigma(\alpha)$, which must be a root of $p(x)$ because

$$0 = \sigma(0) = \sigma(p(\alpha)) = p(\sigma(\alpha)) \quad \text{since } \sigma \text{ is an isomorphism.}$$

Thus, $\sigma(\alpha) = $ some $\alpha_i$, in which case $\sigma = \sigma_i$.

We will call the isomorphisms of a field $E$ into $\mathbb{C}$ its **embeddings** in $\mathbb{C}$.

We are interested in the field $F = \mathbb{Q}(\beta)$ generated by an element $\beta \in E = \mathbb{Q}(\alpha)$, in which case we have fields $E \supseteq F \supseteq \mathbb{Q}$. One thing we already know about this situation is the Dedekind product theorem of Section 6.2, which tells us that the relative dimensions satisfy

$$[E : \mathbb{Q}] = [E : F][F : \mathbb{Q}].$$

An immediate consequence is that the degree $[F : \mathbb{Q}]$ of $\beta$ divides the degree $[E : \mathbb{Q}] = n$ of $\alpha$. More generally, the conjugates $\beta$ are related to the values $\sigma_i(\beta)$ as follows:

**Conjugates of $\beta \in \mathbb{Q}(\alpha)$.** *The values $\sigma_i(\beta)$ are the $[F : \mathbb{Q}]$ conjugates of $\beta$, each occurring $[E : F]$ times.*

*Proof.* Suppose $q(x)$ is the minimal polynomial for $\beta$, with degree $m = [F : \mathbb{Q}]$. Then we know that there are $m$ embeddings $\tau_j : F \to \mathbb{C}$ by Section 4.4. Next observe that each of these embeddings $\tau_j$ can be *extended* to an embedding $\sigma : E \to \mathbb{C}$ by viewing $E$ as a vector space over $F$.

In fact, viewing $E$ as $F(\gamma)$, by the theorem of the primitive element, we see there are exactly $n/m$ such extensions of each $\tau_j$. This is because $\gamma$ is of degree $n/m$, by the Dedekind product theorem, so $\gamma$ can be sent to any of its $n/m$ conjugates by an extension of $\tau_j$.

This gives $n$ embeddings $E \to \mathbb{C}$, which must be the $n$ embeddings $\sigma_i$. It also shows that each of the $m$ distinct values $\tau_j(\beta)$ occurs $[E : F] = n/m$ times among the $n$ values $\sigma_i$. $\qquad\square$

**Corollary.** *The minimal polynomial $q(x)$ of $\beta$ satisfies*

$$q(x)^{[E:F]} = (x - \sigma_1(\beta)) \cdots (x - \sigma_n(\beta)).$$

*Proof.* Each side is a monic polynomial of degree $n$ with roots $\sigma_1(\beta), \ldots, \sigma_n(\beta)$. $\qquad\square$

## Exercises

1. Show, by finding the roots of $x^3 = \alpha^3$ where $\alpha = 2^{1/3}$, that the conjugates of $2^{1/3}$ are $2^{1/3}\zeta$ and $2^{1/3}\zeta^2$, where

$$\zeta = \frac{-1 + \sqrt{-3}}{2}.$$

2. Deduce that the general element $a + b \cdot 2^{1/3} + c \cdot 2^{2/3}$ of $\mathbb{Q}(2^{1/3})$ has the conjugates

$$a + b \cdot 2^{1/3}\zeta + c \cdot 2^{2/3}\zeta^2 \quad \text{and} \quad a + b \cdot 2^{1/3}\zeta^2 + c \cdot 2^{2/3}\zeta$$

3. Show that the conjugates of $\sqrt{2} + \sqrt{3}$ are

$$\sqrt{2} - \sqrt{3}, \quad -\sqrt{2} + \sqrt{3}, \quad -\sqrt{2} - \sqrt{3}$$

   by showing that all these numbers satisfy the same minimal polynomial.

4. Hence, find the conjugates of the general element $a + b\sqrt{2} + c\sqrt{3} + d\sqrt{6}$ of $\mathbb{Q}(\sqrt{2}, \sqrt{3})$.

## 6.7 Discussion

Linear algebra is an ancient subject that was not taken seriously until modern times. It originated as a discipline for solving linear equations and, since linear equations are the easiest to solve, linear algebra remained on the bottom rung of algebra as long as its main aim was thought to be solving equations. It was only when mathematicians turned their attention to *structures*, such as rings, fields, and vector spaces, that linear algebra was seen in a new light: as a necessary foundation for many parts of algebra, geometry, and analysis.

### 6.7.1 Linear Equations

The first important discovery in linear algebra was the algorithm for solving systems of linear equations we call "Gaussian elimination." It was invented not by Gauss, but by Chinese mathematicians about 2000 years ago. Precisely when, we do not know, but it appears in the *Nine Chapters of Mathematical Art*, a famous compilation that was required reading for examination candidates during the Han dynasty (from around 200 BCE to 200 CE).

As most readers will know, "Gaussian elimination" displays a set of $n$ equations in $n$ unknowns in a square array, such as

$$x + y + z = 1$$
$$2x + y + 3z = 0$$
$$x - y + 2z = 4,$$

and subtracts multiples of equations from one another to reduce the system to *triangular form*. Here we remove the $x$ term from the last two equations by subtracting suitable multiples of the first equation from them, obtaining

$$x + y + z = 1$$
$$-y + z = -2$$
$$-2y + z = 3.$$

Then we subtract twice the second equation from the third, obtaining the triangular system

$$x + y + z = 1$$
$$-y + z = -2$$
$$-z = 7.$$

Finally, we obtain the values of $z, y, x$ in that order by working from the bottom up: $z = -7$, hence $y = -5$, hence $x = 13$.

The reduction to triangular form involves only the matrix of coefficients

$$\begin{pmatrix} 1 & 1 & 1 & 1 \\ 2 & 1 & 3 & 0 \\ 1 & -1 & 2 & 4 \end{pmatrix},$$

and indeed the Chinese carried out the process in exactly this way, representing the matrix of numbers by counters on a board.

Thus, it has been clear since ancient times how to solve $n$ linear equations in $n$ unknowns. A more sophisticated question is whether there is a *formula* expressing the unknowns in terms of the coefficients and known operations. For small values of $n$ such formulas were found by the Japanese mathematician Seki in 1683 and Leibniz in 1693. By 1708 Seki had found the underlying concept of **determinant** and defined it inductively. In Europe the general determinant concept was rediscovered by Cramer (1750) and used by him to give a general formula for the solution of linear equations. These nearly simultaneous discoveries underline the "rightness" of the determinant concept for the theory of linear equations.

In the next chapter we will see that the determinant is indeed the "right" concept for explaining solvability of systems of linear equations. However, we will also see that the determinant is quite a complicated object. Thus, Cramer's formula sent linear algebra down the difficult path of determinant theory, diverting it for more than a century from the simpler concepts of linear independence, basis, and dimension.

### 6.7.2 Linear Spaces

As the term "dimension" suggests, linear algebra can be viewed geometrically, particularly when the base field is $\mathbb{R}$. The one-dimensional $\mathbb{R}$-vector space is the line, the two-dimensional $\mathbb{R}$-vector space is the plane, and so on. This idea was generalized to arbitrary $n$ by Grassmann (1844). In one stroke he created an *n-dimensional geometry* in the form of the $n$-dimensional $\mathbb{R}$-vector space. The idea of vector space was already novel, and Grassmann loaded it with the extra novelties of inner and outer products (of which the determinant was merely a special case). Add to this an obscure style of exposition, and the result was a book that Grassmann's contemporaries were completely unable to understand.

Such was the difficult birth of modern linear algebra. Grassmann (1862) made a second attempt, stripping down his idea to the minimum needed for $n$-dimensional Euclidean geometry: the vector space $\mathbb{R}^n$ of real $n$-tuples with the inner product

Figure 6.1 Seki Takakazu (1642–1708) and Gottfried Wilhelm Leibniz (1646–1716). Courtesy of the Japan Academy and licensed under Creative Commons Attribution-NonCommercial-NoDerivatives 4.0 International License (CC BY-NC-ND 4.0), respectively.

Figure 6.2 Hermann Grassmann (1809–1877) and Giuseppe Peano (1858–1932). Licensed under Creative Commons Attribution-ShareAlike 4.0 International License.

$$(x_1, x_2, \ldots, x_n) \cdot (y_1, y_2, \ldots, y_n) = x_1 y_1 + x_2 y_2 + \cdots + x_n y_n.$$

Grassmann even pointed out that the inner product is essentially equivalent to the Pythagorean theorem, but it didn't help. His style was such that few mathematicians had the patience to read him. So his ideas – which included even basics such as invariance of dimension – languished until Peano (who also rescued Grassmann's ideas on arithmetic) published axioms for vector spaces in Peano (1888).

Peano's vector space axioms were the same as those we use today, except that he specialized to the field $F = \mathbb{R}$. The idea of varying the field grew from the work of Dedekind, in the Supplements he wrote for his editions of Dirichlet in 1871, 1879, and 1894. In Dedekind's vector spaces $F$ can be any algebraic number field. The fully fledged notion of $F$-vector space had to wait until the concept of field itself became established in full generality, which happened in the paper of Steinitz (1910).

### 6.7.3 Linear Maps

In algebra today it is commonplace to say that we study not only certain kinds of structure – such as rings, fields, or vector spaces – but also the *structure-preserving maps* between them: their **homomorphisms**. However, this is a fairly recent realization. In the evolution of an algebraic concept, the appropriate notion of homomorphism is often last to emerge, if only because we need to know precisely what the structure *is* before we can talk about preserving it.

This is certainly how the concept of **linear map** emerged. Particular linear maps have been around for a long time; for example, in the form of *substitutions* or *changes of variable*, like those used by Lagrange to study quadratic forms (see Section 3.3). But linear maps were not considered as

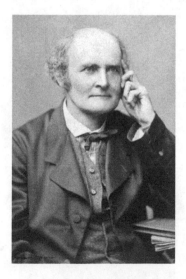

Figure 6.3 Arthur Cayley (1821–1895). Licensed under Creative Commons Attribution-ShareAlike 4.0 International License.

mathematical objects until the 19th century, when the matrix notation of Cayley (1855) gave the object a symbolic existence.

Matrices quickly took on a life of their own, as Cayley moved to consider the *sum* and *product* of matrices and the matrix *inverse*. As Katz and Parshall (2014), p. 358, wrote

> Cayley's paper represented a significant step in the evolution of algebra, for it showed how matrices could be understood as *constructs*, subject to many of the usual operations of algebra.

But matrices are not the last word on linear maps. The matrix for a linear map of a vector space $V$ depends on the choice of basis for $V$ and, ideally, one would like to discuss linear maps without choosing a basis. The *definition* of linear map in Section 6.3 suggests that basis-free linear algebra is possible. Modern books, starting with Halmos (1942), have indeed taken this approach.

However, there is still something to be said for matrices:

> We share a philosophy about linear algebra: we think basis-free, we write basis-free, but when the chips are down we close the office door and compute with matrices like fury.
>
> *(Irving Kaplansky, writing of himself and Halmos in Kaplansky (1991)).*

# 7

# Determinant Theory

## Preview

The determinant function is a concept of linear algebra frequently skimmed over or minimized in modern treatments of the subject. At the same time, books on algebraic number theory tend to assume sophisticated properties of the determinant – such as its relationship to **trace**, **norm**, and **characteristic polynomial** – to be already known from a basic linear algebra course.

Under these circumstances it seems useful to develop the determinant concept from scratch and then transition to its applications in algebraic number fields. This is what we aim to do in this chapter, hopefully making the book more self-contained without greatly increasing its size.

We begin with an elementary treatment of determinants, due to Artin (1942). His approach leads rapidly to methods for evaluating determinants, applications to linear equations, and to the all-important **multiplicative property**.

We can then prove the **invariance** of the determinant under change of basis and deduce the basis-independence of trace, norm, and characteristic polynomial. This leads in turn to relations between the trace and norm of an algebraic number and the roots of its minimal polynomial.

With these foundations we can then introduce the **discriminant**, which tests whether an $n$-tuple of members of a field $F$ of degree $n$ over $\mathbb{Q}$ is a basis for $F$. This paves the way for the study of **integral bases** in the next chapter. These generalize the concept of basis from vector spaces to certain kinds of modules, such as the algebraic number rings $\mathbb{Z}_E$.

Figure 7.1 Emil Artin (1898–1962) (used with permission of Tom Artin and his siblings).

## 7.1 Axioms for the Determinant

There are many ways to define the determinant function and derive its basic properties, none of them completely straightforward. One that is elementary, yet algebraic in spirit, is given in Artin (1942), pages 12–20. We follow Artin's approach in this section and the next, except that we assume only that matrix entries are from a ring $R$, because only ring operations are involved. In Section 7.3 we specialize to matrices with entries in a field or a domain.

The first step is to impose certain requirements, or *axioms*, on a function $D(A)$ of $n \times n$ matrices $A$ that we expect the determinant to have. The axioms are chosen to be as simple as possible, yet capable (as we will see) of implying all properties of the determinant function. To state the axioms, we let $A_1, \ldots, A_n$ denote the columns of $A$, and view $D$ as a function $D(A_1, \ldots, A_n)$ of its columns. We write $D(\ldots, A_k, \ldots)$ or $D(\ldots, A_k, A_{k+1}, \ldots)$ when $D$ is viewed as a function of $A_k$, or of $A_k$ and $A_{k+1}$, with the other columns held constant.

**Axiom 1.** $D$ is linear on each column $A_k$; that is,

$$D(\ldots, A_k + A'_k, \ldots) = D(\ldots, A_k, \ldots) + D(\ldots, A'_k, \ldots)$$

and

$$D(\ldots, cA_k, \ldots) = cD(\ldots, A_k, \ldots) \quad \text{for each } c \in R.$$

**Axiom 2.** $D(\ldots, A_k, A_{k+1}, \ldots) = 0$ when $A_k = A_{k+1}$.

**Axiom 3.** $D(I) = 1$, where $I$ is the $n \times n$ identity matrix. In terms of columns, $D(U_1, \ldots, U_n) = 1$, where $U_1, \ldots, U_n$ are the columns of $I$.

For the moment we are interested only in what *follows* from these axioms. When enough consequences are known, we will know exactly what $D$ looks like, and its existence will be easy to prove.

One after another, we derive the following consequences.

1. Putting $c = 0$ in Axiom 1, we get $D = 0$ when any column is zero.
2. By Axiom 1, if we add $cA_{k+1}$ to column $k$ we get

$$D(\ldots, A_k + cA_{k+1}, A_{k+1}, \ldots) = D(\ldots, A_k, A_{k+1}, \ldots)$$
$$+ cD(\ldots, A_{k+1}, A_{k+1}, \ldots)$$
$$= D(\ldots, A_k, A_{k+1}, \ldots),$$

because $D(\ldots, A_{k+1}, A_{k+1}, \ldots) = 0$ by Axiom 2. Thus, the value of $D$ is unchanged if we add a multiple of one column to another.
3. By adding or subtracting (take $c = -1$) one column from another, we can make the following changes in columns $k$ and $k + 1$.

$$A_k, A_{k+1} \mapsto A_k, A_{k+1} + A_k \qquad \text{adding column } k \text{ to column } k + 1,$$
$$\mapsto -A_{k+1}, A_{k+1} + A_k \quad \text{subtracting column } k + 1 \text{ from column } k,$$
$$\mapsto -A_{k+1}, A_k \qquad \text{adding column } k \text{ to column } k + 1.$$

Thus, $D$ is unaltered by this change, by consequence 2.
4. Multiplying the new column $k$ by $-1$, which reverses the sign of $D$ by Axiom 1, we obtain $A_{k+1}, A_k$ in columns $k$ and $k + 1$. Thus, swapping adjacent columns reverses the sign of $D$.
5. Any two columns $A_k$ and $A_l$ can be made adjacent by a sequence of swaps of adjacent columns. So if $A_k = A_l$, then $D = 0$ by Axiom 2.
6. By adjacent swaps and applying Axiom 2 as in consequences 2 and 3, we can similarly show $D$ is unchanged by adding a multiple of *any* column to another, and $D$'s sign is reversed by swapping *any* two columns.
7. Any permutation $A_{\sigma(1)}, \ldots, A_{\sigma(n)}$ of the columns $A_1, \ldots, A_n$ is obtainable by a sequence of swaps, so

$$D(A_{\sigma(1)}, \ldots, A_{\sigma(n)}) = \pm D(A_1, \ldots, A_n).$$

In particular, by Axiom 3,

$$D(U_{\sigma(1)}, \ldots, U_{\sigma(n)}) = \pm 1, \quad \text{since} \quad D(U_1, \ldots, U_n) = 1.$$

With these facts we can compute the effect on $D(A_1, \ldots, A_n)$ of replacing each $A_k$ by an arbitrary linear combination of columns:

$$A'_k = b_{1k}A_1 + \cdots + b_{nk}A_n, \quad \text{where } b_{1k}, \ldots, b_{nk} \in R.$$

The idea is to expand $D(A'_1, \ldots, A'_n)$ into a sum of terms $\pm D(A_1, \ldots, A_n)$, each multiplied by a product of constants $b_{ij}$, using Axiom 1 to break up $A'_1, A'_2, \ldots$ in turn. At first we get terms $D(A_{i_1}, \ldots, A_{i_n})$ with arbitrary subscripts $i_k$. Those with two equal subscripts vanish by consequence 5; the others get a $\pm$ sign depending on the swaps needed to put the subscripts in the order $1, 2, \ldots, n$.

The process can be illustrated in the case $n = 2$, where

$$
\begin{aligned}
D(A'_1, A'_2) &= D(b_{11}A_1 + b_{12}A_2, A'_2) \\
&= b_{11}D(A_1, A'_2) + b_{12}D(A_2, A'_2) \\
&= b_{11}D(A_1, b_{21}A_1 + b_{22}A_2) + b_{12}(A_2, b_{21}A_1 + b_{22}A_2) \\
&= b_{11}b_{21}D(A_1, A_1) + b_{11}b_{22}D(A_1, A_2) \\
&\quad + b_{12}b_{21}D(A_2, A_1) + b_{12}b_{22}D(A_2, A_2) \\
&= b_{11}b_{22}D(A_1, A_2) + b_{12}b_{21}D(A_2, A_1) \\
&= b_{11}b_{22}D(A_1, A_2) - b_{12}b_{21}D(A_1, A_2) \\
&= (b_{11}b_{22} - b_{12}b_{21})D(A_1, A_2).
\end{aligned}
$$

In the general case $D(A'_1, \ldots, A'_n)$ is the sum of terms

$$
\pm b_{1\sigma(1)}b_{2\sigma(2)} \cdots b_{n\sigma(n)} D(A_{\sigma(1)}, \ldots, A_{\sigma(n)}),
$$

over all **permutations** $\sigma$ of $1, 2, \ldots, n$, because $D$ is zero on any sequence of $n$ columns $A_i$ with a repeated column. It follows, by swapping columns $A_i$ until they are in order, that

$$
D(A'_1, \ldots, A'_n) = \sum_\sigma \pm D(A_1, \ldots, A_n)b_{1\sigma(1)}b_{2\sigma(2)} \cdots b_{n\sigma(n)}, \qquad (*)
$$

in which each sign ($+$ or $-$) is determined by the permutation $\sigma$.

The formula $(*)$ was derived from Axioms 1 and 2 alone. If we now take $A$ to be the identity matrix and assume Axiom 3, then $D(A_1, \ldots, A_n) = 1$ and also $A'_k = B_k$, the $k$th column of the matrix $B$ of the coefficients $b_{ij}$. This gives a formula for $D$ of an arbitrary matrix $B$:

$$
D(B_1, \ldots, B_n) = \sum_\sigma \pm b_{1\sigma(1)}b_{2\sigma(2)} \cdots b_{n\sigma(n)}, \qquad (**)
$$

So, *if $D$ exists, it is uniquely determined* by Axioms 1, 2, and 3.

In the light of the formula $(**)$, formula $(*)$ becomes

$$
D(A'_1, \ldots, A'_n) = D(A_1, \ldots, A_n)D(B_1, \ldots, B_n). \qquad (***)
$$

Thus $D$ has the **multiplicative property**, $D(AB) = D(A)D(B)$, since the matrix $A'$ with columns $A'_k$ is none other than the product of the matrices $A$

and $B$: the equations expressing the $A'_k$ in terms of $A_1, \ldots, A_n$ specifically give

$$j\text{th entry in column } A'_k = b_{1k}a_{j1} + b_{2k}a_{j2} + \cdots + b_{nk}a_{jn},$$

which is the product of the $j$th row of $A$ with the $k$th column of $B$.

## Exercises

The terms $\pm D(A_1, \ldots, A_n)b_{1\sigma(1)}b_{2\sigma(2)} \cdots b_{n\sigma(n)}$ whose sum is $D(A'_1, \ldots, A'_n)$ can be described in a little more detail.

1. Prove by induction on $n$ that, for each permutation $\sigma$ of $1, \ldots, n$, the term $b_{1\sigma(1)}b_{2\sigma(2)} \cdots b_{n\sigma(n)}$ occurs exactly once in the expansion of $D(A'_1, \ldots, A'_n)$.

It is of particular interest to see how the sign of $b_{1\sigma(1)}b_{2\sigma(2)} \cdots b_{n\sigma(n)}$ depends on $\sigma$. To do this we distinguish between "even" and "odd" permutations $\sigma$. A permutation $\sigma$ is called **even** if it has an even number of inversions; that is, an even number of pairs $i < j$ such that $\sigma(i) > \sigma(j)$. Otherwise, $\sigma$ is called **odd**.

2. Why is the identity permutation $\sigma(i) = i$ even?
3. Show that interchange of $\sigma(i)$ and $\sigma(j)$—a **transposition**—changes the permutation $\sigma$ from even to odd or else from odd to even.
4. Show that any permutation is obtainable from the identity permutation by a finite sequence of transpositions.
5. Conclude that $\sigma$ is even if and only if it is a product of an even number of transpositions.
6. Show that the sign of $b_{1\sigma(1)}b_{2\sigma(2)} \cdots b_{n\sigma(n)}$ is $+$ if $\sigma$ is even and $-$ if not.

## 7.2 Existence of the Determinant Function

If $A$ is an $n \times n$ matrix, we prove existence (and hence uniqueness) of a function $D(A)$ satisfying Axioms 1, 2, and 3 by induction on $n$. For $n = 1$ we have $A = (a_{11})$, and it is easy to check that $D(A) = a_{11}$ satisfies Axioms 1, 2, and 3.

Now suppose that $D$ exists for all $(n - 1) \times (n - 1)$ matrices and consider an $n \times n$ matrix $A$, with $(i, j)$ entry $a_{ij}$. Define the **cofactor** $A_{ij}$ of $a_{ij}$ to be $D$ of the $(n - 1) \times (n - 1)$ submatrix of $A$ obtained by deleting row $i$ and column $j$, with a $\pm$ sign determined by the checkerboard pattern

$$\begin{pmatrix} + & - & + & \cdots \\ - & + & - & \cdots \\ + & - & + & \cdots \\ \vdots & \vdots & \vdots & \end{pmatrix}$$

For any $i$ among $1, \ldots, n$ consider the function

$$D(A) = a_{i1}A_{i1} + a_{i2}A_{i2} + \cdots + a_{in}A_{in},$$

viewed as a function $D(A_1, \ldots, A_n)$ of the columns of $A$.

**Determinant properties of $D$.** *$D$ satisfies Axioms 1, 2, and 3; hence it is the unique determinant function and is independent of $i$.*

*Proof.* If $A_k$ is replaced by $cA_k$, then $a_{ik}$ is replaced by $ca_{ik}$, but $A_{ik}$ is unchanged because it does not involve the column $A_k$. The remaining $A_{ij}$ *do* involve $A_k$, hence they are multiplied by $c$, while $a_{ij}$ is not. Hence,

$$D(\ldots, cA_k, \ldots) = cD(\ldots, A_k, \ldots).$$

A similar argument shows that

$$D(\ldots, A_k + A_k', \ldots) = D(\ldots, A_k, \ldots) + D(\ldots, A_k', \ldots),$$

so $D$ satisfies Axiom 1.

Next, if $A_k = A_{k+1}$, then $A_{ik}$ and $A_{i,k+1}$ are $D$ of the same matrix but with opposite signs, while the other $A_{ij}$ have two equal columns and hence are zero by consequence 5 in the previous section. Thus, $D = 0$ when $A_k = A_{k+1}$, so Axiom 2 is satisfied.

Finally, if $A$ is the $n \times n$ identity matrix, then $a_{ii} = 1$ and its cofactor $A_{ii}$ is $D$ of the $(n - 1) \times (n - 1)$ identity matrix, hence $A_{ii} = 1$ by induction. Every other $a_{ij} = 0$, so $D(A) = 1$ and Axiom 3 is satisfied.                            $\square$

### 7.2.1 Further Rules for Computing Determinants

From now on we will call $D$ by its usual name det, for **determinant**.

The formula (**) in Section 7.1 in principle gives a rule for computing det, though it is not very practical unless many entries are zero or if one wants only a particular term, such as the product $b_{11}b_{22} \cdots b_{nn}$ of the diagonal elements. The latter is actually the case when we study the characteristic polynomial in Section 7.4.

The above definition of det via cofactors,

$$\det(A) = a_{i1}A_{i1} + a_{i2}A_{i2} + \cdots + a_{in}A_{in},$$

gives a way to compute det inductively by **expansion on rows**, as it is called. It reduces computation of $\det(A)$ to computing the determinants $A_{ij}$ of smaller matrices, and hence ultimately to det of $1 \times 1$ matrices.

Anything done with rows can also be done with columns, because the formula (**) in Section 7.1 shows that interchange of rows and columns makes no change in the value of det. Namely, since the $+$ or $-$ sign depends only on the permutation $\sigma$,

$$D(B_1, \ldots, B_n) = \sum_{\sigma} \pm b_{1\sigma(1)} b_{2\sigma(2)} \cdots b_{n\sigma(n)} = \sum_{\sigma} \pm b_{\sigma(1)1} b_{\sigma(2)2} \cdots b_{\sigma(n)n},$$

because the second sum is simply a rearrangement of the first, putting the second subscripts in increasing order instead of the first subscripts.

It follows in particular that **expansion on column** $i$ is valid:

$$\det(A) = a_{1i} A_{1i} + a_{2i} A_{2i} + \cdots + a_{ni} A_{ni}.$$

It also follows that the matrix $A^t$ obtained by interchanging rows and columns of $A$—the **transpose** of $A$—has the same determinant as $A$.

It must be confessed that all of these methods are hopelessly inefficient in practice, but they suffice for our theoretical purposes. Practical methods for computing determinants are related to practical methods for solving systems of linear equations, which readers have probably seen in introductory linear algebra courses. (In particular, we can mimic "Gaussian elimination" to reduce the computation of the determinant to that of finding the determinant of a triangular matrix, which is simply the product of its diagonal elements.)

## Exercises

First, a couple of interesting particular computations:

1. Prove that

$$\det \begin{pmatrix} a & b & c \\ b & c & a \\ c & a & b \end{pmatrix} = 3abc - a^3 - b^3 - c^3.$$

2. Prove that

$$\det \begin{pmatrix} 1 & a & a^2 \\ 1 & b & b^2 \\ 1 & c & c^2 \end{pmatrix} = (a - b)(b - c)(c - a)$$

by expansion on the first column. (We take up a generalization of this result in Section 7.6.)

3. How could one anticipate the factorization $(a - b)(b - c)(c - a)$ *without* expanding the determinant?

Next, a useful general fact about multiplying row elements by the "wrong" cofactors. Suppose that $A$ is the matrix with entries $a_{ij}$ and let $A'$ be the result of replacing row $j$ of $A$ by row $i$.

4. Explain why $\det(A') = a_{i1} A_{j1} + \cdots + a_{in} A_{jn}$, where the $A_{jk}$ are the cofactors of $a_{jk}$ in $A$.
5. Deduce that $a_{i1} A_{j1} + \cdots + a_{in} A_{jn} = 0$ when $j \neq i$.
6. State and prove the corresponding fact about multiplying elements of a column by the "wrong" cofactors.

## 7.3  Determinants and Linear Equations

The results of the last two sections, including the rules for calculating det and its multiplicative property, are valid for matrices with entries in any ring. In the present section we specialize to matrices with entries from a field $F$, because we wish to study linear dependence. Thus, the columns $A_1, \ldots, A_n$ of an $n \times n$ matrix $A$ are now vectors in the $n$-dimensional space $F^n$.

Under these conditions we have the following criterion for $\det(A) = 0$.

**Criterion for zero determinant.** *If $A$ is an $n \times n$ matrix with entries from a field $F$, then $\det(A) = 0$ if and only if the columns $A_1, \ldots, A_n$ of $A$ are linearly dependent.*

*Proof.* Suppose first that $A_1, \ldots, A_n$ are linearly dependent. Then some column $A_k$ is a linear combination $C_k$ of the others, in which case

$$\det(\ldots, A_k, \ldots) = \det(\ldots, A_k - C_k, \ldots) = \det(\ldots, 0, \ldots) = 0$$

by the consequences of Axioms 1, 2, and 3 in Section 7.1.

To prove the converse, first observe what happens if every vector $V \in F^n$ is a linear combination of $A_1, \ldots, A_n$. Then, in particular,

$$U_i = b_{1i} A_i + \cdots + b_{ni} A_n, \quad \text{for } i = 1, \ldots, n,$$

where the $U_i$ are the columns of the identity matrix $I$. In that case,

$$1 = \det I = \det(A) \det(B), \quad \text{where } B \text{ is the matrix of the } b_{ij},$$

by the multiplicative property of det. Hence, $\det(A) \neq 0$.

Now suppose $A_1, \ldots, A_n$ are linearly independent and consider any vector $V \in F^n$. The $n + 1$ vectors $A_1, \ldots, A_n, V \in F^n$ are linearly dependent by the exchange lemma of Section 6.2, so there are $x_1, \ldots, x_n, y \in F$, not all zero, such that

$$A_1 x_1 + \cdots + A_n x_n + V y = \mathbf{0}.$$

In particular, $y \neq 0$, otherwise $A_1, \ldots, A_n$ are linearly dependent. This means we can write an arbitrary $V$ as a linear combination of $A_1, \ldots, A_n$:

$$V = -\frac{1}{y}\left(A_1 x_1 + \cdots + A_n x_n\right).$$

Then, as we have just shown, $\det(A) \neq 0$.                                  □

**Corollary.** *A set of linear homogeneous equations with matrix $A$*

$$a_{i1} x_1 + \cdots + a_{in} x_n = 0, \quad \text{for } i = 1, \ldots, n, \qquad (*)$$

*has a nonzero solution if and only if* $\det(A) = 0$.

*Proof.* This set of equations is equivalent to the vector equation

$$A_1 x_1 + \cdots + A_n x_n = \mathbf{0},$$

where $A_1, \ldots, A_n$ are the columns of $A$. So the existence of a nonzero solution is equivalent to linear dependence of $A_1, \ldots, A_n$, and hence to $\det(A) = 0$.                                  □

This corollary is the property of det assumed in Section 4.6, in order to prove that algebraic integers are closed under sum, difference, and product. We do not need more general applications of determinants to linear equations, but in view of their general importance, we will look at them in the exercises below.

**Remark.** The corollary extends to equations with coefficients in a *domain $R$*, (which was the case in Section 4.6) by embedding $R$ in its field of fractions in order to carry out the argument about linear independence. So the conclusion that $\det(A) = 0$, where $A$ is the matrix of coefficients, remains valid.

## Exercises

We now generalize the equations (*) in the above corollary to the general set of $n$ linear equations in $n$ unknowns,

$$a_{i1} x_1 + \cdots + a_{in} x_n = b_i, \quad \text{for } i = 1, \ldots, n, \qquad (**)$$

This set is equivalent to the vector equation

$$A_1x_1 + \cdots + A_nx_n = B, \quad \text{where } B = \begin{pmatrix} b_1 \\ \vdots \\ b_n \end{pmatrix},$$

and $A_1, \ldots, A_n$ are the columns of the matrix $A$ with entries $a_{ij}$.

1. Explain why there is a solution of (**) for any $B$ if and only if $\det(A) \neq 0$.

Now we will solve the equations (**) by choosing any $k$ and multiplying the $i$th equation by the cofactor $A_{ik}$ of $a_{ik}$, so that it becomes

$$a_{i1}A_{ik}x_1 + \cdots + a_{in}A_{ik}x_n = b_iA_{ik}.$$

2. Summing the latter equations for $i = 1, \ldots, n$, show that

$$\text{coefficient of } x_k = a_{1k}A_{1k} + \cdots + a_{nk}A_{nk} = \det(A).$$

3. Also show, with the help of exercise 5 in Section 7.2, that

$$\text{coefficient of } x_j = a_{1j}A_{1k} + \cdots + a_{nj}A_{nk} = 0, \quad \text{when } j \neq k.$$

4. Deduce that $\det(A)x_k = b_1A_{1k} + \cdots + b_nA_{nk}$, and that the right-hand side is the determinant of the matrix $A'$ obtained by replacing the $k$th column of $A$ by $B$.

The formula for solving the equations (**) implicit in exercise 4 is called **Cramer's rule**, after its appearance in Cramer (1750).

Figure 7.2  Gabriel Cramer (1704–1752). Courtesy of the Bibliotèque de Genève.

## 7.4 Basis Independence

If a linear map $f$ has a matrix $M$ with respect to a certain basis, and if $B$ is the matrix for the linear map sending the old basis to a new one, then the matrix for $f$ with respect to the new basis is $BMB^{-1}$. (Reading from right to left, $BMB^{-1}$ says "map the new basis to the old one, do $M$, then map the old basis back to the new.")

A function $g$ of $M$ is said to be **basis-independent**, and hence dependent only on the underlying linear map $f$, if $g(BMB^{-1}) = g(M)$ for any basis change matrix $B$. An example of basis independence that we have already seen, in the exercises to Section 6.3, is that of the det function. We revisit this proof now.

**Basis independence of det.** *If $B$ is a basis change matrix, then*

$$\det(BMB^{-1}) = \det(B).$$

*Proof.* Since $BB^{-1} = I$, taking det of both sides and using the multiplicative property gives

$$1 = \det(I) = \det(BB^{-1}) = \det(B)\det(B^{-1}).$$

This implies $\det(B^{-1}) = \det(B)^{-1}$, and it follows that

$$\det(BMB^{-1}) = \det(B)\det(M)\det(B^{-1})$$
$$= \det(B)\det(B)^{-1}\det(M) = \det(M). \qquad \square$$

A variation of this argument gives basis independence of the **characteristic polynomial** of a matrix $M$, charpoly($M$), which is defined to be $\det(xI - M)$, where $I$ is the identity matrix.

**Basis independence of charpoly.** *If $B$ is a basis change matrix,*

$$charpoly(BMB^{-1}) = charpoly(M).$$

*Proof.* When $B$ is the matrix for a change of basis, notice that

$$xI - BMB^{-1} = xBIB^{-1} - BMB^{-1} = B(xI - M)B^{-1},$$

since $BIB^{-1} = BB^{-1}I = I$. It follows that

$$\det(xI - BMB^{-1}) = \det(B(xI - M)B^{-1})$$
$$= \det(B)\det(xI - M)\det(B^{-1})$$
$$= \det(B)\det(B)^{-1}\det(xI - M)$$
$$= \det(xI - M).$$

That is,

$$\text{charpoly}(BMB^{-1}) = \text{charpoly}(M). \qquad \qquad \square$$

We make use of this basis independence in the next section, where we calculate the characteristic polynomial for the "multiply by $\alpha$" map in terms of the most convenient basis for a given algebraic number $\alpha$. But first, let us pursue an obvious corollary of the invariance of charpoly:

**Corollary.** *All coefficients of* charpoly$(M)$ *are basis-independent.*

One coefficient of interest is the constant term, which equals $(-1)^n \det(M)$ if $M$ is an $n \times n$ matrix. This can be seen by setting $x = 0$. Another is the coefficient of $x^{n-1}$. To find this coefficient we let

$$M = \begin{pmatrix} a_{11} & a_{12} & \cdots & a_{1n} \\ a_{21} & a_{22} & \cdots & a_{2n} \\ \vdots & & & \vdots \\ a_{n1} & a_{n2} & \cdots & a_{nn} \end{pmatrix}, \quad \text{so}$$

$$xI - M = \begin{pmatrix} x - a_{11} & -a_{12} & \cdots & -a_{1n} \\ -a_{21} & x - a_{22} & \cdots & -a_{2n} \\ \vdots & & & \vdots \\ -a_{n1} & -a_{n2} & \cdots & x - a_{nn} \end{pmatrix}.$$

Then it is clear that all the $x^{n-1}$ terms in $\det(xI - M)$ arise from the product of the diagonal terms, $(x - a_{11})(x - a_{22}) \cdots (x - a_{nn})$, and hence

$$\text{coefficient of } x^{n-1} = -(a_{11} + a_{22} + \cdots + a_{nn}).$$

**Definition and notation.** The **trace** of an $n \times n$ matrix $M$ with $(i, j)$-entry $a_{ij}$ is defined by

$$\text{Tr}(M) = a_{11} + a_{22} + \cdots + a_{nn}.$$

With this definition we can now write

$$\text{charpoly}(M) = x^n - \text{Tr}(M)x^{n-1} + \cdots + (-1)^n \det(M).$$

**Remark.** One property of trace that follows immediately from its definition is $\text{Tr}(M + M') = \text{Tr}(M) + \text{Tr}(M')$.

## Exercises

The basis-independence of trace can also be proved directly from its definition.

1. Prove from the definition of trace that $\text{Tr}(CD) = \text{Tr}(DC)$.
2. Deduce from the previous exercise that $\text{Tr}(A) = \text{Tr}(BAB^{-1})$.

## 7.5 Trace and Norm of an Algebraic Number

Given an algebraic number field $E = \mathbb{Q}(\alpha)$, we now adapt the idea of Section 6.4 to express any $\beta \in \mathbb{Q}(\alpha)$ as a matrix: namely, the matrix $\boldsymbol{\beta}$ that expresses the linear map $x \mapsto \beta x$ ("multiplication by $\beta$") of $\mathbb{Q}(\alpha)$ relative to the basis elements $1, \alpha, \ldots, \alpha^{n-1}$.

The matrix $\boldsymbol{\beta}$ for multiplication by $\beta$ of course depends on the choice of basis, but two important functions of a matrix are basis independent: its *trace* and its *determinant*, as we saw in the previous section. Because of this, the trace and determinant of the matrix for multiplication by $\beta$ depend only on $\beta$ and the field $E$.

**Definitions and notation.** If $\beta \in E = \mathbb{Q}(\alpha)$, we define its **trace** in the extension $E$ of $\mathbb{Q}$, $\text{Tr}_{E/\mathbb{Q}}(\beta)$, to be $\text{Tr}(\boldsymbol{\beta})$, where $\boldsymbol{\beta}$ is the matrix for multiplication by $\beta$ in $E$. $N_{E/\mathbb{Q}}(\beta)$, the **norm of** $\beta$ in the extension $E$, is defined to be $\det(\boldsymbol{\beta})$.[1]

When it is clear which extension field $E$ of $\mathbb{Q}$ we mean, we simply write $\text{Tr}(\beta)$ and $N(\beta)$. We know from the sections above that trace and det are basis independent, and also their main properties, such as:

$$\text{Tr}(M + M') = \text{Tr}(M) + \text{Tr}(M'), \quad \det(MM') = \det(M)\det(M'),$$

which imply

$$\text{Tr}(\beta + \beta') = \text{Tr}(\beta) + \text{Tr}(\beta'), \quad N(\beta\beta') = N(\beta)N(\beta'), \quad \text{for any } \beta, \beta' \in E.$$

The latter **multiplicative property of norm** follows from the multiplicative property of det and is a broad generalization of the multiplicative property of norm we observed in the special case of Gaussian integers back in Section 2.3. The multiplicative property is also related to the minimal polynomial for the number $\beta$, and its relation to the matrix $\boldsymbol{\beta}$ and its characteristic polynomial.

---

[1] The subscript $E/\mathbb{Q}$ does *not* denote a quotient structure, but is rather the algebraist's pun for "$E$ over $\mathbb{Q}$." Much as I deplore the pun, and its clash with the notation for quotient structures, it seems too widely used to resist.

**Minimal and characteristic polynomials.** *If $p(x)$, of degree $n$, is the minimal polynomial of $\alpha$, and $\boldsymbol{\alpha}$ is the matrix for the map $u \mapsto \alpha u$ of $\mathbb{Q}(\alpha)$ with respect to the basis $1, \alpha, \ldots, \alpha^{n-1}$, then the characteristic polynomial of $\boldsymbol{\alpha}$ equals $p(x)$.*

*Proof.* Suppose that $p(x) = x^n + a_{n-1}x^{n-1} + \cdots + a_0$, where $a_0, \ldots, a_{n-1} \in \mathbb{Q}$. Then, as we saw in Section 6.4,

$$
\boldsymbol{\alpha} = \begin{pmatrix}
0 & 1 & 0 & \cdots & 0 & 0 \\
0 & 0 & 1 & \cdots & 0 & 0 \\
\vdots & \vdots & \vdots & & \vdots & \vdots \\
0 & 0 & 0 & \cdots & 0 & 1 \\
-a_0 & -a_1 & -a_2 & \cdots & -a_{n-2} & -a_{n-1}
\end{pmatrix}.
$$

It follows that

$$
xI - \boldsymbol{\alpha} = \begin{pmatrix}
x & -1 & 0 & \cdots & 0 & 0 \\
0 & x & -1 & \cdots & 0 & 0 \\
\vdots & \vdots & \vdots & & \vdots & \vdots \\
0 & 0 & 0 & \cdots & x & -1 \\
a_0 & a_1 & a_2 & \cdots & a_{n-2} & x + a_{n-1}
\end{pmatrix},
$$

By expanding along the first column, repeatedly, we find that

$$
\det(xI - \boldsymbol{\alpha}) = x^n + a_{n-1}x^{n-1} + \cdots + a_0 = p(x). \qquad \square
$$

We generalize this result to the minimal polynomial $q(x)$ for an arbitrary $\beta \in \mathbb{Q}(\alpha)$ by following the idea in Section 6.6. In outline, the argument is this.

Let $\boldsymbol{\beta}^*$ be the matrix for the map $v \mapsto \beta v$ in $\mathbb{Q}(\beta)$ with respect to the basis $1, \beta, \ldots, \beta^{m-1}$, where $m$ is the degree of $q(x)$. Expand this basis to a basis of $\mathbb{Q}(\alpha)$ by multiplying each $\beta^j$ by a basis element $\gamma^i$ for $\mathbb{Q}(\alpha)$ over $\mathbb{Q}(\beta)$, where $\gamma$ exists by the theorem of the primitive element. Then the matrix for $v \mapsto \beta v$ in $\mathbb{Q}(\alpha)$ becomes

$$
\boldsymbol{\beta} = \begin{pmatrix}
\boldsymbol{\beta}^* & \mathbf{0} & \cdots & \mathbf{0} \\
\mathbf{0} & \boldsymbol{\beta}^* & \cdots & \mathbf{0} \\
\vdots & & & \vdots \\
\mathbf{0} & \mathbf{0} & \cdots & \boldsymbol{\beta}^*
\end{pmatrix},
$$

where $\mathbf{0}$ denotes the $m \times m$ zero matrix, when the basis vectors are ordered

$$
1, \beta, \ldots, \beta^{m-1}; \quad \gamma, \gamma\beta, \ldots, \gamma\beta^{m-1}; \quad \cdots \; ;
$$
$$
\gamma^{n/m-1}, \gamma^{n/m-1}\beta, \ldots, \gamma^{n/m-1}\beta^{m-1}.
$$

It follows from the argument for $\alpha$ above that the characteristic polynomial of $\beta^*$ is the minimal polynomial of $\beta$, namely $q(x)$. Then the shape of $\beta$ shows that its characteristic polynomial is $q(x)^{n/m}$, which equals $q(x)^{[E:F]}$.

Now $q(x)^{[E:F]}$ is the polynomial whose factorization we found in Section 6.6:

$$q(x)^{[E:F]} = (x - \sigma_1(\beta)) \cdots (x - \sigma_n(\beta)). \qquad (*)$$

This formula gives us formulas for $\mathrm{Tr}(\beta)$ and $N(\beta)$:

**Trace and norm in terms of the $\sigma_i$.** *For any $\beta \in F$, where $F \supseteq \mathbb{Q}$ is a field extension of degree $n$,*

$$Tr(\beta) = \sigma_1(\beta) + \cdots + \sigma_n(\beta), \quad N(\beta) = \sigma_1(\beta) \cdots \sigma_n(\beta),$$

*where the $\sigma_i$ are the $n$ embeddings $F \to \mathbb{C}$.*

*Proof.* We know from the previous section that the characteristic polynomial of an $n \times n$ matrix $M$ is related to $\mathrm{Tr}(M)$ and $\det(M) = N(M)$ by

$$\mathrm{charpoly}(M) = x^n - \mathrm{Tr}(M)x^{n-1} + \cdots + (-1)^n \det(M).$$

Comparing this with the expression (*) for charpoly($\beta$), we see that

$$\mathrm{Tr}(\beta) = \sigma_1(\beta) + \cdots + \sigma_n(\beta), \quad N(\beta) = \sigma_1(\beta) \cdots \sigma_n(\beta). \qquad \square$$

**Remarks.** Since the matrix $\beta$ for multiplication by $\beta$ expresses a linear map of $\mathbb{Q}(\alpha)$ (viewed as a vector space over $\mathbb{Q}$), its entries are rational numbers. It follows that its trace and norm—that is, $\mathrm{Tr}(\beta)$ and $N(\beta)$—are rational numbers. Indeed, if $\beta$ is an algebraic integer, then the entries $-a_0, \ldots, -a_{n-1}$ in the matrix $\beta$ are ordinary integers, and hence $\mathrm{Tr}(\beta)$ and $N(\beta)$ are ordinary integers.

The formula $N(\beta) = \sigma_1(\beta) \cdots \sigma_n(\beta)$ confirms the agreement between the determinant concept of norm and the special norms used in Sections 1.5, 2.3, and 2.9. There we defined the norm of $\beta$ to be the product of the conjugates of $\beta$, and found that it was rational-valued and multiplicative. In those cases there were two conjugates, but in the case of degree $n$ the conjugates of $\beta$ are $\sigma_1(\beta), \ldots, \sigma_n(\beta)$.

This formula also shows the multiplicative property of norm – which we first saw as a consequence of the multiplicative property of det – in the light of the multiplicative property of the isomorphisms $\sigma_i$. Namely,

$$N(\beta\beta') = \sigma_1(\beta\beta') \cdots \sigma_n(\beta\beta') = \sigma_1(\beta)\sigma_1(\beta') \cdots \sigma_n(\beta)\sigma_n(\beta') = N(\beta)N(\beta').$$

## Exercises

A field in which traces and norms are easily computed from conjugates is the **cyclotomic field** $\mathbb{Q}(\zeta_p)$, where $p$ is prime and

$$\zeta_p = \cos\frac{2\pi}{p} + i\sin\frac{2\pi}{p}.$$

A cyclotomic field we already know is $\mathbb{Q}(\sqrt{-3})$, generated by $\zeta_3 = \frac{-1+\sqrt{-3}}{2}$ (which we sometimes called just $\zeta$). We know that $\mathbb{Q}(\zeta_3)$ is of degree 2 over $\mathbb{Q}$, and in fact $\mathbb{Q}(\zeta_p)$ is of degree $p-1$, since its minimal polynomial is

$$x^{p-1} + x^{p-2} + \cdots + x + 1, \tag{*}$$

the nontrivial factor of the polynomial $x^p - 1$ obviously satisfied by $\zeta_p$. We proved irreducibility of the polynomial (*) in the exercises to Section 4.7.

Now we will look at some consequences of irreducibility for conjugates, trace, and norm in $\mathbb{Q}(\zeta_p)$.

1. Explain why $\zeta_p, \zeta_p^2, \ldots, \zeta_p^{p-1}$ are the conjugates of $\zeta_p$.
2. Find the sum of these conjugates, and hence show that

$$\mathrm{Tr}(\zeta_p) = \mathrm{Tr}(\zeta_p^2) = \cdots = \mathrm{Tr}(\zeta_p^{p-1}) = -1.$$

3. Explain why $\mathrm{Tr}(1) = p - 1$ and deduce that

$$\mathrm{Tr}(1 - \zeta_p) = \mathrm{Tr}(1 - \zeta_p^2) = \cdots = \mathrm{Tr}(1 - \zeta_p^{p-1}) = p.$$

4. Also explain why

$$N(1 - \zeta_p) = (1 - \zeta_p)(1 - \zeta_p^2) \cdots (1 - \zeta_p^{p-1}).$$

Now we can compute $N(1 - \zeta_p)$ in a different way as the constant term of its minimal polynomial.

5. Explain why the minimal polynomial for $x = 1 - \zeta_p$ is

$$(1 - x)^{p-1} + (1 - x)^{p-2} + \cdots + (1 - x) + 1$$

and deduce that $p = N(1 - \zeta_p) = (1 - \zeta_p)(1 - \zeta_p^2) \cdots (1 - \zeta_p^{p-1}).$

In the exercises to Section 8.3 these calculations will give the *integers* of $\mathbb{Q}(\zeta_p)$.

## 7.6 Discriminant

The theorem of the primitive element tells us that we are lucky in the theory of algebraic number fields: each of them is a vector space over $\mathbb{Q}$ generated by a single element $\alpha$ and with basis $1, \alpha, \ldots, \alpha^{n-1}$. For *rings* of algebraic *integers*, coming in the next chapter, the situation is not so rosy. Even for the integers in algebraic number fields of degree $n$, there is not generally a single generator, nor a convenient basis. To prepare ourselves to face these difficulties, we make a study of bases in general and, in particular, look for a way of testing whether a given $n$-tuple of members of a field $F$ of degree $n$ is a basis of $F$.

The tool for answering this question is the **discriminant**.

**Definition.** The **discriminant** $\Delta(\omega_1, \ldots, \omega_n)$ of $\omega_1, \ldots, \omega_n \in F$, where $F \supseteq \mathbb{Q}$ is an extension of degree $n$, is defined to be $\det(M)^2$, where

$$M = \begin{pmatrix} \sigma_1(\omega_1) & \sigma_1(\omega_2) & \cdots & \sigma_1(\omega_n) \\ \sigma_2(\omega_1) & \sigma_2(\omega_2) & \cdots & \sigma_2(\omega_n) \\ \vdots & & & \vdots \\ \sigma_n(\omega_1) & \sigma_n(\omega_2) & \cdots & \sigma_n(\omega_n) \end{pmatrix},$$

and the $\sigma_i : F \to \mathbb{C}$ are the $n$ embeddings of $F$.

We will find that $\det(M)$ "discriminates" between bases and nonbases accordingly as it is nonzero or zero. The reason for squaring is that $\det(M)$ can change sign for a trivial reason – exchanging two of the $\sigma_i$ – so we want to be able to ignore its sign.

Our first result gives an alternative formula for $\Delta(\omega_1, \ldots, \omega_n)$, as det of the **trace matrix** $T$ whose $(i, j)$-entry is $T_{ij} = \text{Tr}_{F/\mathbb{Q}}(\omega_i \omega_j)$. As usual, we will write simply Tr for $\text{Tr}_{F/\mathbb{Q}}$.

**Discriminant in terms of trace.** *If we set $T_{ij} = Tr(\omega_i \omega_j)$ and let $T$ be the matrix with $(i, j)$-entry $T_{ij}$, then*

$$\Delta(\omega_1, \ldots, \omega_n) = \det(T).$$

*Proof.* By definition, $\Delta(\omega_1, \ldots, \omega_n) = \det(M)^2$, where $M$ is the matrix with $(i, j)$-entry $\sigma_i(\omega_j)$. From Section 7.2 we know that the same determinant belongs to the **transpose** $M^t$, whose $(i, j)$-entry is $\sigma_j(\omega_i)$. Therefore, by the multiplicative property of det, $\det(M)^2$ is the determinant of $M^t M$, whose $(i, j)$-entry is

$\sigma_1(\omega_i)\sigma_1(\omega_j) + \cdots + \sigma_n(\omega_i)\sigma_n(\omega_j)$

$= \sigma_1(\omega_i\omega_j) + \cdots + \sigma_n(\omega_i\omega_j)$     because $\sigma_1, \ldots, \sigma_n$ are isomorphisms

$= \text{Tr}(\omega_i\omega_j)$         by the formula for trace in the previous section

$= T_{ij}$                                     $\square$

An example where we can explicitly calculate the trace is in the field $\mathbb{Q}(\alpha)$, where $\alpha$ is an algebraic number of degree $n$. In this case we can use the standard basis $1, \alpha, \ldots, \alpha^{n-1}$, so $\omega_j = \alpha^{j-1}$. Then if we let $\sigma_i(\alpha) = \alpha_i$, the definition of discriminant says $\Delta(1, \alpha, \ldots, \alpha^{n-1})$ *is the square of*

$$\det \begin{pmatrix} 1 & \alpha_1 & \alpha_1^2 & \cdots & \alpha_1^{n-1} \\ 1 & \alpha_2 & \alpha_2^2 & \cdots & \alpha_2^{n-1} \\ \vdots & & & & \vdots \\ 1 & \alpha_n & \alpha_n^2 & \cdots & \alpha_n^{n-1} \end{pmatrix}$$

This is a classical determinant that goes back to the 18th century, called the **Vandermonde determinant**.[2] It can be evaluated with the help of the following remarks. We first suppose that $\alpha_1, \ldots, \alpha_n$ are simply distinct "variables," or "indeterminates."

1. By summing the degrees of the columns, $0 + 1 + 2 + \cdots + (n-1) = n(n-1)/2$, we see that the determinant is a polynomial in the variables $\alpha_1, \ldots, \alpha_n$, each term of which has total degree $n(n-1)/2$.
2. The determinant becomes zero if any $\alpha_i = \alpha_j$ for $i \neq j$, because two rows become equal. So the determinant has a factor $(\alpha_j - \alpha_i)$ for each pair $i, j$ with $i < j$, by the corollary in Section 1.8 (the factor theorem for multivariable polynomials).
3. There are $n(n-1)/2$ such pairs. Namely, they are half of the pairs with $i \neq j$, for which we can choose $i$ in $n$ ways and then $j$ in $n-1$ ways.
4. Therefore, the Vandermonde determinant equals some constant times the product $\prod_{i<j}(\alpha_j - \alpha_i)$

By looking at a specific term in the determinant – say, the product of the diagonal elements – we can show that the constant equals 1. What is more important is that *we are interpreting $\alpha_1, \ldots, \alpha_n$ as the conjugates* of $\alpha$, so they are all different. Therefore,

$$\prod_{i<j}(\alpha_i - \alpha_j), \text{ and hence } \Delta(1, \alpha, \ldots, \alpha^{n-1}), \text{ is nonzero.}$$

We now leverage this discovery into a proof that the discriminant of *any* basis is nonzero, with the help of the following theorem.

**Discriminant of a transformed basis.** *If $\omega_1, \ldots, \omega_n$ is a basis for a field $E$ of degree $n$ over $\mathbb{Q}$, and $\omega_1', \ldots, \omega_n' \in E$, then*

$$\Delta(\omega_1', \ldots, \omega_n') = \det(A)^2 \Delta(\omega_1, \ldots, \omega_n),$$

*where $A$ is the matrix of the linear map sending the $\omega_i$ to the $\omega_i'$.*

---

[2] Although Vandermonde wrote a paper about determinants in 1771, its connection with this particular determinant is not straightforward. See Ycart (2013) for the whole story.

*Proof.* Since $\omega_1, \ldots, \omega_n$ is a basis, we can express the $\omega_i'$ by linear equations

$$\omega_i' = a_{1i}\omega_1 + \cdots + a_{ni}\omega_n. \tag{*}$$

Let $A$ be the matrix with $(i, j)$-entry $a_{ij}$. Now $\Delta(\omega_1', \ldots, \omega_n')$ is the square of the det of

$$M' = \begin{pmatrix} \sigma_1(\omega_1') & \sigma_1(\omega_2') & \cdots & \sigma_1(\omega_n') \\ \sigma_2(\omega_1') & \sigma_2(\omega_2') & \cdots & \sigma_2(\omega_n') \\ \vdots & & & \vdots \\ \sigma_n(\omega_1') & \sigma_n(\omega_2') & \cdots & \sigma_n(\omega_n') \end{pmatrix},$$

and

$$\sigma_k(\omega_i') = a_{1i}\sigma_k(\omega_1) + \cdots + a_{ni}\sigma_k(\omega_n)$$

by (*), since $\sigma_k$ is an isomorphism.

It follows that $M' = AM$, where $M$ is the matrix whose squared det equals $\Delta(\omega_1, \ldots, \omega_n)$. Taking det of both sides of $M' = BM$, and squaring, therefore gives

$$\Delta(\omega_1', \ldots, \omega_n') = \det(A)^2 \Delta(\omega_1, \ldots, \omega_n) \qquad \square$$

**Corollary.** *For any* $\omega_1, \ldots, \omega_n \in E$, *a field of degree n over* $\mathbb{Q}$,

$$\omega_1, \ldots, \omega_n \text{ is a basis of } E \Leftrightarrow \Delta(\omega_1, \ldots, \omega_n) \neq 0.$$

*Proof.* The theorem of the primitive element gives a basis of $E$ of the form $1, \alpha, \ldots, \alpha^{n-1}$, for which $\Delta \neq 0$. The matrix $A$ of the map sending this basis to any other basis is invertible, and hence $\det(A) \neq 0$. Then the theorem above gives $\Delta \neq 0$ for the new basis.

Conversely, if $A$ maps a basis to a *non*basis, then $\det(A) = 0$, so $\Delta$ of the nonbasis is zero. $\qquad \square$

**Remark.** The formula for the discriminant of a transformed basis is a big generalization of the discovery of Lagrange, mentioned in Section 3.3, about the discriminants of quadratic forms. The maps considered by Lagrange have determinant $\pm 1$, which is why they leave the discriminant unchanged.

## Exercises

1. Show from the formula for $\Delta$ that $\Delta(1, i) = -4$, which agrees with Lagrange's discriminant $D$ for the form $x^2 + y^2 = (x + iy)(x - iy)$.

2. Likewise, show that $\Delta\left(1, \sqrt{-2}\right) = -8$, which agrees with Lagrange's discriminant $D$ for the form $x^2 + 2y^2 = (x + \sqrt{-2}y)(x - \sqrt{-2}y)$.

3. Show that $\Delta(1, \zeta) = -3$, which agrees with Lagrange's discriminant $D$ for the form $x^2 + xy + y^2 = (x - \zeta y)(x - \bar{\zeta} y)$, where $\zeta = \frac{-1 + \sqrt{-3}}{2}$.

For a more general result that encompasses the three above, consider a quadratic form $ax^2 + bxy + cy^2$ for which Lagrange's discriminant $D = b^2 - 4ac$ is not a square. Then $ax^2 + bxy + cy^2$ does not have a rational factorization, and the numbers

$$\frac{-b + \sqrt{D}}{2a} \quad \text{and} \quad \frac{-b - \sqrt{D}}{2a}$$

are conjugate members of the quadratic field $\mathbb{Q}(\sqrt{D})$.

4. Show that

$$ax^2 + bxy + cy^2 = \left( x - \frac{-b + \sqrt{D}}{2a} y \right) \left( x - \frac{-b - \sqrt{D}}{2a} y \right)$$

5. Explain why $1, \frac{-b + \sqrt{D}}{2}$ is a basis for the quadratic field $\mathbb{Q}(\sqrt{D})$, and show that

$$\Delta \left( 1, \frac{-b + \sqrt{D}}{2} \right) = D.$$

The reason for choosing $\frac{-b + \sqrt{D}}{2}$ as a basis element, rather than $\frac{-b + \sqrt{D}}{2a}$, is that $\frac{-b + \sqrt{D}}{2}$ is an *algebraic integer*, as defined in Section 4.6.

6. Check that $\frac{-b + \sqrt{D}}{2}$ is the root of a monic polynomial equation with ordinary integer coefficients.

Thus we have actually found an **integral basis** for the field, an idea that will be important in the next chapter – see Section 8.3.

## 7.7 Discussion

The theory of norm and discriminant in this chapter is essentially the same as that given by Dedekind in the 1870s; see (Dedekind, 1996, pp. 112–116). He derives the multiplicative property of the norm directly from the multiplicative property of the isomorphisms $\sigma_i$ but uses determinant theory to prove the discriminant criterion for a basis. Dedekind shows that the discriminant of any basis is nonzero via the Vandermonde determinant, whose value is nonzero "by virtue of a well known proposition in the theory of determinants" (1996, p. 112). Thus, Dedekind assumed that his readers had a strong a background in determinant theory.

### 7.7.1 History of Determinant Theory

As mentioned in Section 6.7, determinant theory was once the main topic in linear algebra. It attracted the attention of leading mathematicians and was one of the main branches of algebra in the 19th century. For this reason, it became the subject of one the most monumental histories of any branch of mathematics, the four volumes of Sir Thomas Muir's *The Theory of Determinants in the Historical Order of Development*, written between 1890 and the 1920s. Muir's book was reprinted as Muir (1960), but in recent decades it has dropped out of sight, no doubt because linear algebra is now universal and determinant theory is (relatively) marginal. But still, it *is* important, as we have tried to show in this chapter, so let us review some of its history.

A noticable feature of the history of determinants, as Muir seldom failed to point out, is that its main results have often been discovered and rediscovered independently. Determinants themselves were discovered independently by Seki in Japan in 1683 and Leibniz in Europe at approximately the same time. Seki's early results were for small determinants (up to $5 \times 5$), but by 1710 he had found an inductive definition of $n \times n$ determinants via cofactors— essentially the one used in Section 7.2. This is also known as the Laplace expansion because, yes, it was rediscovered by Laplace (1772). Leibniz's work on determinants, which was never published, has been little known until recently. The eminent Leibniz scholar Eberhard Knobloch has concluded, in Knobloch (2013), that Leibniz found all the basic results, including the Laplace expansion, the sign of permutations, and Cramer's rule, by 1684.

As mentioned in Section 6.7, Cramer first published the determinant rule for solving systems of linear equations. The rule appeared in the widely read book by Cramer (1750) about algebraic curves, so by 1750 the European mathematical community was finally aware of the determinant concept. Its basic properties were elucidated over the next 50–60 years:

- Vandermonde (1771) sketched the foundations of the theory.
- Laplace (1772) (re)discovered the cofactor expansion.
- Rothe (1800) defined even and odd permutations.
- In 1801 Gauss introduced the term "determinant" and discovered the multiplicative property.

• In 1812 Binet and Cauchy independently discovered the multiplicative property, and Cauchy gave a satisfactory proof.

Summing up these developments from the viewpoint of the 1890s, Muir (1960) wrote in his volume 1 (p. 130):

> It is not too much to say, though it may come to many as a surprise, that the ordinary textbooks of determinants supplied to university students of the present day do not contain much more of the general theory than is to be found in Cauchy's memoir of about eighty years ago. ... It is no doubt impossible to call him, as some have done, the formal founder of the theory. This honour is certainly due to Vandermonde, who, however, erected on the foundation comparatively little of the superstructure. ... Cauchy relaid the foundation, rebuilt the whole, and initiated new enlargements; the result being an edifice which the architects of today may still admire and find worthy of study.

The modern theory begins with Grassmann (1844) and his sophisticated concept of outer product. Muir actually notes Grassmann's contribution, but he does not recognize the rising tide of the modern linear algebra that was eventually to submerge his beloved theory of determinants.

It is therefore interesting that the decidedly modern Artin (1942) took the trouble to present an elementary treatment of determinants, which was not really needed for the rest of his book. Perhaps he felt that determinants were being neglected, or handled inelegantly, and he wanted to take the opportunity to educate his readers.

# 8

# Modules

## Preview

We now consider the effect of replacing the field $F$ in the definition of vector space by a ring $R$. The resulting structures are called $R$-**modules**, and they have exactly the same definition as $F$-vector spaces, except that the "scalars" come from the ring $R$ instead of from the field $F$. And just as the structure of algebraic number fields can be clarified by viewing them as $\mathbb{Q}$-vector spaces, the structure of algebraic integer rings can be clarified by viewing them as $\mathbb{Z}$-modules.

However, this idea meets difficulties because modules do not always have the standard vector space features, such as basis and dimension. In general, $R$-modules do *not* have bases, but there is an important class of those that do: the so-called **free $R$-modules**. For these, we can describe linear maps by matrices, as we do for vector spaces, and prove that there is an invariant dimension, at least for free modules with a finite basis.

However, finding bases for modules requires deeper linear algebra than it does for vector spaces. There is no "primitive element theorem" saying that an algebraic number ring is $\mathbb{Z}[\alpha]$ for some algebraic integer $\alpha$. And proving that submodules of a free $R$-module $M$ have dimension no greater than that of $M$ requires a restriction on $R$: that it be a principal ideal domain. Fortunately, this applies to $R = \mathbb{Z}$, so we escape the difficulties in the case that originally motivated the theory of modules.

In particular, the algebraic integer rings $\mathbb{Z}_E$ are finitely generated $\mathbb{Z}$-modules, and we can find bases for them with the help of the discriminant.

With these results, we can revisit rings and ideals from the viewpoint of modules, which is a convenient generalization of both.

171

## 8.1 From Vector Spaces to Modules

As mentioned in Section 6.3, a module is the analogue of a vector space where
the field $F$ is replaced by a ring $R$.

**Definition.** An $R$-**module**, is a set $M$ of objects $u, v, w, \ldots$ which can be added
and multiplied by members of $R$.

Addition in an $R$-module has the properties of an **Abelian group**, namely

$$u + v = v + u,$$
$$u + (v + w) = (u + v) + w,$$
$$u + 0 = u,$$
$$u + (-u) = 0.$$

Multiplication, by any member of $R$, has the properties

$$a(bu) = (ab)u,$$
$$1u = u,$$
$$a(u + v) = au + av,$$
$$(a + b)u = au + bu.$$

Thus, calculations in a module, like those in a vector space, resemble
calculations with numbers, except that multiplication is restricted to members
of $R$ and there is not generally any division. A vector space is the special
case of an $R$-module where $R$ is a field, but there are many other important
examples. Here are some.

1. Any ring $R$ is itself an $R$-module. It is **generated** by the element 1, in the
   sense that $R = \{r1 : r \in R\}$. If $R$ is a subring of a ring $S$, then $S$ is an
   $R$-module, generated by 1 and certain elements of $S$ that are not in $R$.
2. Any Abelian group $G$ is a $\mathbb{Z}$-module. Writing the group operation as $+$
   (as in Section 3.7), if $g \in G$ and $n \in \mathbb{Z}$ we let $ng = g + \cdots + g$ (with $n$
   summands) when $n > 0$, and the obvious things when $n \leq 0$.
3. The ring $\mathbb{Z}[\sqrt{2}] = \{a + b\sqrt{2} : a, b \in \mathbb{Z}\}$ is a $\mathbb{Z}$-module, generated by the
   two elements 1 and $\sqrt{2}$.
4. Similarly, the ring $\mathbb{Z}[\sqrt{-5}]$ is a $\mathbb{Z}$-module, generated by the two elements
   1 and $\sqrt{-5}$. More interestingly, the *ideal*

$$\left\{ 2m + (1 + \sqrt{-5})n : m, n \in \mathbb{Z} \right\}$$

in $\mathbb{Z}[\sqrt{-5}]$ is a $\mathbb{Z}$-module, generated by the elements 2 and $1 + \sqrt{-5}$.

5. Any ideal $I \subseteq$ ring $R$ is an $R$-module $\subseteq R$ (an $R$-**submodule**). Conversely, any $R$-submodule of $R$ is an ideal because the sum of two elements of $I$ is in $I$, and the product of an element of $I$ by an element of $R$ is in $I$.

Because of the last two examples, modules have an important role to play in the study of ideals. One of our aims will be to prove that the integers of an algebraic number field constitute a finitely generated $\mathbb{Z}$-module, and that the same is true of their ideals.

One of the properties of ideals that extends to modules is the ability to form *quotients*.

**Definitions and Notation.** If $M'$ is an $R$-submodule of an $R$-module $M$, then $a, b \in M$ are said to be **congruent mod** $M'$, written as

$$a \equiv b \bmod M',$$

if $a - b \in M'$. The class $[a] = \{b : a \equiv b \bmod M'\}$ is called the **congruence class** of $a$ mod $M'$, and the congruence classes mod $M'$ constitute the **quotient module** $M/M'$.

Of course, to justify these definitions, we have to show that congruence mod $M'$ is an equivalence relation, and check that the congruence classes form an $R$-module when addition and scalar multiplication are defined by

$$[a] + [b] = [a + b], \quad c[a] = [ca], \quad \text{for any } c \in R.$$

These facts can be verified by much the same arguments that we used for the special case of congruence mod $n$ in $\mathbb{Z}$ (Section 1.6). In fact, it seems that modules got their name (from Dedekind) because they admit a relation of congruence *modulo M*.

To some extent, we can study $R$-modules by methods of linear algebra. But certain convenient properties of vector spaces, such as the existence of bases, do not hold for all modules. Because of that, we often study $R$-modules for restricted classes of rings $R$, such as the principal ideal domains (PIDs) mentioned in Section 5.2. Since $\mathbb{Z}$ is a PID, we can exploit this property in rings that are $\mathbb{Z}$-modules, even when they are not PIDs themselves.

The definition of linear map for $R$-modules is exactly the same as that for vector spaces, but with "scalars" from the ring $R$ instead of the field $F$. We may also call it an $R$-linear map and represent it by a matrix with entries in $R$. However, this is possible only if the $R$-module has a *basis*, which we investigate in the next section when $R = \mathbb{Z}$.

## Exercises

Consider the $\mathbb{Z}$-module $\mathbb{Z}[\sqrt{-5}]$ and the $\mathbb{Z}$-submodules (which are ideals) introduced in Section 5.1 and its exercises:

$$I = \left\{ 2m + (1 + \sqrt{-5})n : m, n \in \mathbb{Z} \right\},$$

$$J = \left\{ 3m + (1 + \sqrt{-5})n : m, n \in \mathbb{Z} \right\}$$

We saw in the exercises to Section 5.3 that $I$ and $J$ have different quotients $\mathbb{Z}[\sqrt{-5}]/I = \mathbb{F}_2$ and $\mathbb{Z}[\sqrt{-5}]/J = \mathbb{F}_3$. Nevertheless, $I$ and $J$ are geometrically similar – they have the *same shape*.

1. By viewing the pictures of $I$ and $J$ in Section 5.1, show that, among several elements of equal size,

   - 2 and $1 - \sqrt{-5}$ (in that order) are the smallest two members of $I$,
   - $1 + \sqrt{-5}$ and 3 (in that order) are the smallest two members of $J$.

2. Show that $\frac{2}{1 - \sqrt{-5}} = \frac{1 + \sqrt{-5}}{3}$.
3. Deduce from this equation that both $I$ and $J$ may be viewed as the vertex sets of parallelograms of the *same shape*; that is, with the same angles and the same ratio (length of longer side)/(length of shorter side).
4. Show in fact that $J$ may be obtained from $I$ by magnifying by 3/2 and rotating by a certain angle.
5. Explain why the *principal* ideals of $\mathbb{Z}[\sqrt{-5}]$ all have the same shape, which is different from the shape of $I$ and $J$.

These two "shapes" of ideals correspond to what are called *ideal classes*, which happen to be members of the **class group** briefly mentioned in Section 3.4.

## 8.2 Algebraic Number Fields and Their Integers

As we saw in Section 4.6, the algebraic integers in an algebraic number field $E$ form a ring because their sums and products are integers, also in $E$. We denote this ring by $\mathbb{Z}_E$ and view it as a $\mathbb{Z}$-module by interpreting $n\alpha$, for $n \in \mathbb{Z}$ and $\alpha \in \mathbb{Z}_E$, as the $|n|$-fold sum

$$\alpha + \cdots + \alpha \text{ if } n > 0, \quad (-\alpha) + \cdots + (-\alpha) \text{ if } n < 0, \quad \text{and} \quad 0 \text{ if } n = 0.$$

The members of $E$ are "fractions" formed from the integers of $E$, in fact they are fractions whose denominators are ordinary integers.

**Algebraic numbers as fractions.** *If the field $E$ is an extension of $\mathbb{Q}$ of finite degree, then each $\beta \in E$ is of the form $\alpha/b$, where $\alpha \in \mathbb{Z}_E$ and $b \in \mathbb{Z}$.*

*Proof.* If $E$ is an extension of degree $m$, then each $\beta \in E$ satisfies an equation of the form

$$a_m x^m + a_{m-1} x^{m-1} + \cdots + a_0 = 0, \text{ where } a_0, \ldots, a_m \in \mathbb{Z}.$$

Multiplying through by $a_m^{m-1}$, we obtain the equation

$$(a_m x)^m + a_{m-1}(a_m x)^{m-1} + \cdots + a_0 a_m^{m-1} = 0,$$

also satisfied by $x = \beta$. But the latter equation is a monic equation in $a_m x$, so $a_m \beta$ is an *algebraic integer* $\alpha \in \mathbb{Z}_E$.

Thus, $\beta = \alpha/a_m$ is the quotient of $\alpha \in \mathbb{Z}_E$ by an ordinary integer $a_m \in \mathbb{Z}$. $\square$

A simple instance of the theorem is the extension field $\mathbb{Q}(\sqrt{2})$ of $\mathbb{Q}$, whose integers are the numbers $p + q\sqrt{2}$ where $p, q \in \mathbb{Z}$. We have seen that a general element of $\mathbb{Q}(\sqrt{2})$ is of the form $a + b\sqrt{2}$, where $a, b \in \mathbb{Q}$. And $a + b\sqrt{2}$ is of the form $(p + q\sqrt{2})/r$, where $p, q, r \in \mathbb{Z}$ and $r$ is a common denominator of $a$ and $b$.

Now for some corollaries to the representation of algebraic numbers as fractions.

**Corollary 1.** $E = \text{Frac}(\mathbb{Z}_E)$.

*Proof.* Because $\text{Frac}(\mathbb{Z}_E)$ includes all fractions of the form $\alpha/b$, with $\alpha \in \mathbb{Z}_E$ and $b \in \mathbb{Z} \subseteq \mathbb{Z}_E$, and these fractions make up all of $E$. $\square$

**Corollary 2.** If $E$ has degree $m$ over $\mathbb{Q}$, then $E$ has a basis of $m$ elements of $\mathbb{Z}_E$ as a vector space over $\mathbb{Q}$.

*Proof.* We know from Section 4.5 that $E$ of degree $m$ is a vector space of dimension $m$ over $\mathbb{Q}$, so it has a basis consisting of $m$ elements $\beta_i \in E$. By the theorem above, each $\beta_i = \alpha_i/b_i$, where $\alpha_i \in \mathbb{Z}_E$ and $b_i \in \mathbb{Z}$.

If we multiply the basis elements $\beta_i$ by the least common multiple $b$ of the $b_i$ we obtain elements $\omega_i = b\beta_i$ which belong to $\mathbb{Z}_E$ and which obviously still form a basis for $E$ as a vector space over $\mathbb{Q}$. $\square$

The second corollary means that each $\beta \in E$ can be written in the form

$$\beta = r_1 \omega_1 + \cdots + r_m \omega_m, \text{ where } r_1, \ldots, r_m \in \mathbb{Q} \text{ and } \omega_1, \ldots, \omega_m \in \mathbb{Z}_E.$$

If in fact $r_1, \ldots, r_m \in \mathbb{Z}$, then $\beta \in \mathbb{Z}_E$, by the ring properties of $\mathbb{Z}_E$, but the converse does not necessarily hold. If it does, the $\omega_i$ form what is called an **integral basis** for $\mathbb{Z}_E$, which is something much to be desired.

## Exercises

We saw, in exercise 1 of Section 4.6, that the ring of integers of $\mathbb{Q}$ is $\mathbb{Z}$. In the case of quadratic fields $\mathbb{Q}(\sqrt{d})$ for an ordinary integer $d$, it is quite easy to describe the integers of the field:

- For $d \not\equiv 1 \pmod 4$, they are the $a + b\sqrt{d}$ with $a, b \in \mathbb{Z}$.
- For $d \equiv 1 \pmod 4$, they are the $a + b\sqrt{d}$ with $2a, 2b \in \mathbb{Z}$.

1. Show first that we can assume $d$ **squarefree** (not divisible by a square greater than 1), because if $d = m^2 n$, then $\mathbb{Q}(\sqrt{d}) = \mathbb{Q}(\sqrt{n})$.
2. Show that the equation satisfied by $x = a + b\sqrt{d}$, for $a, b \in \mathbb{Q}$, is

$$x^2 + Ax + B = 0, \text{ where } A = -2a \text{ and } B = a^2 - db^2.$$

Thus, for $a + b\sqrt{d}$ to be an integer of $\mathbb{Q}(\sqrt{d})$ we need $A, B \in \mathbb{Z}$. The condition $A = -2a \in \mathbb{Z}$ implies either $a \in \mathbb{Z}$ or $2a \in \mathbb{Z}$, with $2a$ odd. Let us see where both these possibilities lead.

3. Assuming $a \in \mathbb{Z}$ and $a^2 - db^2 \in \mathbb{Z}$, show that $db^2 \in \mathbb{Z}$ and hence $b^2 \in \mathbb{Z}$ because $d$ is squarefree.
4. Conclude from exercise 3 that if $a \in \mathbb{Z}$, then $b \in \mathbb{Z}$.
5. Assuming $2a \in \mathbb{Z}$ with $2a$ odd, and $a^2 - db^2 \in \mathbb{Z}$, show

$$d(2b)^2 \equiv (2a)^2 \equiv 1 \pmod 4.$$

6. Deduce from exercise 5 that $d \equiv 1 \pmod 4$ and $2b \in \mathbb{Z}$ with $2b$ odd.

## 8.3 Integral Bases

**Definition.** If $E \supseteq \mathbb{Q}$ is a field extension of finite degree, then $\omega_1, \ldots, \omega_m \in \mathbb{Z}_E$ are said to be an **integral basis** for $\mathbb{Z}_E$ if each $\beta \in \mathbb{Z}_E$ is uniquely expressible in the form

$$\beta = b_1\omega_1 + \cdots + b_m\omega_m \quad \text{with } b_1, \ldots, b_m \in \mathbb{Z}.$$

We now show that $\mathbb{Z}_E$ always has an integral basis, with the help of the discriminant from Section 7.6.

**Existence of an integral basis.** *If $E \supseteq \mathbb{Q}$ is a field extension of degree $m$, then $\mathbb{Z}_E$ has an integral basis of $m$ elements.*

*Proof.* We have just seen, in Corollary 2, that $E$ has a basis over $\mathbb{Q}$ consisting of $\omega_1, \ldots, \omega_m \in \mathbb{Z}_E$. For such a basis, $\Delta(\omega_1, \ldots, \omega_m) \in \mathbb{Z}$ because $\Delta$ is the determinant of the matrix $T$ with entries $\text{Tr}(\omega_i \omega_j)$ in $\mathbb{Z}$, by the formula for discriminant in terms of trace in Section 7.6 and the remark near the end of Section 7.5. Also, $\Delta(\omega_1, \ldots, \omega_m)$ is positive by its definition in Section 7.6.

This means we can choose a basis $\omega_1, \ldots, \omega_m$ of members of $\mathbb{Z}_E$ whose discriminant is *minimal*. We now show that such a basis is an integral basis for $\mathbb{Z}_E$. If not, there is some $\omega \in \mathbb{Z}_E$ whose (unique) expression

$$\omega = r_1 \omega_1 + \cdots + r_m \omega_m, \text{ with } r_1, \ldots, r_m \in \mathbb{Q},$$

has some $r_i \notin \mathbb{Z}$. By reordering the basis elements, if necessary, we can assume that $r_1 \notin \mathbb{Z}$.

Then, if $a_1$ is the integer nearest to $r_1$, we have $|r_1 - a_1| \leq 1/2$. If we now replace the basis element $\omega_1$ by

$$\omega_1' = \omega - a_1 \omega_1 = (r_1 - a_1)\omega_1 + r_2 \omega_2 + \cdots + r_m \omega_m,$$

then $\omega_1' \in \mathbb{Z}_E$ by the ring properties of $\mathbb{Z}_E$, and we still have a basis of $E$ because

$$\omega_1 = \frac{1}{r_1 - a_1}\left(\omega_1' - r_2 \omega_2 - \cdots - r_m \omega_m\right) \quad \text{and} \quad r_1 - a_1 \neq 0.$$

The matrix for the change of basis from $\omega_1, \omega_2, \ldots, \omega_m$ to $\omega_1', \omega_2, \ldots, \omega_m$ is

$$A = \begin{pmatrix} r_1 - a_1 & r_2 & r_3 & \cdots & r_m \\ 0 & 1 & 0 & \cdots & 0 \\ 0 & 0 & 1 & \cdots & 0 \\ \vdots & & & & \vdots \\ 0 & 0 & 0 & \cdots & 1 \end{pmatrix}.$$

So, by the formula for the discriminant of a transformed basis in Section 7.6,

$$\Delta(\omega_1', \omega_2, \ldots, \omega_m) = \det(A)^2 \Delta(\omega_1, \omega_2 \ldots, \omega_m)$$
$$= (r_1 - a_1)^2 \Delta(\omega_1, \omega_2, \ldots, \omega_m)$$
$$< \Delta(\omega_1, \omega_2, \ldots, \omega_m)$$

because $|r_1 - a_1| \leq 1/2$. This contradicts the minimality of $\Delta(\omega_1, \omega_2 \ldots, \omega_m)$, so $\omega_1, \omega_2, \ldots, \omega_m$ is indeed an integral basis. $\square$

**Corollary.** *Any integral basis of $\mathbb{Z}_E$ has $m$ members.*

*Proof.* Any integral basis of $\mathbb{Z}_E$ is also a vector space basis for $E$ over $\mathbb{Q}$, because each $\beta \in E$ equals $\alpha/b$ where $\alpha \in \mathbb{Z}_E$ and $b \in \mathbb{Z}$. Hence, the basis has $m$ members by the invariance of vector space dimension. $\square$

With this theorem and its corollary, we have clawed back the advantages that vector spaces usually have over modules – basis and dimension – albeit for a limited class of modules. What we have shown is that the ring $\mathbb{Z}_E$ of integers in an algebraic number field of degree $m$ is a **free $\mathbb{Z}$-module of rank $m$**, in terminology to be defined in the next section. Of course, the rings $\mathbb{Z}_E$ are precisely the rings we are interested in, so we could be satisfied with this special case. But it is worth saying a little more about the general concepts of free $R$-modules and their rank.

## Exercises

Following the determination of the integers of $\mathbb{Q}(\sqrt{d})$ for squarefree $d \in \mathbb{Z}$, in the previous exercise set, we can now find an integral basis for these integers.

1. If $d \not\equiv 1 \pmod{4}$, show that $1, \sqrt{d}$ is an integral basis for the integers of $\mathbb{Q}(\sqrt{d})$.

2. If $d \equiv 1 \pmod{4}$, show that $1, \frac{-1+\sqrt{d}}{2}$ is an integral basis for the integers of $\mathbb{Q}(\sqrt{d})$.

These results confirm that the integers we previously found in the fields $\mathbb{Q}(\sqrt{-3})$ and $\mathbb{Q}(\sqrt{-5})$ – namely the numbers $a + b\frac{-1+\sqrt{-3}}{2}$ and $a + b\sqrt{-5}$ for $a, b$ in $\mathbb{Z}$ – are in fact all the integers in these fields.

Another interesting example is the **cyclotomic field** $\mathbb{Q}(\zeta_p)$, where $p$ is prime and $\zeta_p$ is the following solution of $x^p - 1 = 0$:

$$\zeta_p = \cos\frac{2\pi}{p} + i\sin\frac{2\pi}{p}.$$

We calculated some norms and traces for this field in the exercises to Section 7.5, in particular

$$\mathrm{Tr}(1 - \zeta_p) = \mathrm{Tr}(1 - \zeta_p^2) = \cdots = \mathrm{Tr}(1 - \zeta_p^{p-1}) = p, \qquad (*)$$

$$N(1 - \zeta_p) = (1 - \zeta_p)(1 - \zeta_p^2)\cdots(1 - \zeta_p^{p-1}) = p. \qquad (**)$$

First, we show that $(1 - \zeta_p)\mathbb{Z}_{\mathbb{Q}(\zeta_p)} \cap \mathbb{Z} = p\mathbb{Z}$; that is, the multiples of $1 - \zeta_p$ by integers of $\mathbb{Q}(\zeta_p)$ that are in $\mathbb{Z}$ are the multiples of $p$.

3. Deduce from $(**)$ that $p \in (1 - \zeta_p)\mathbb{Z}_{\mathbb{Q}(\zeta_p)}$, and hence that $(1 - \zeta_p)\mathbb{Z}_{\mathbb{Q}(\zeta_p)} \cap \mathbb{Z}$ is an ideal of $\mathbb{Z}$ containing $p\mathbb{Z}$.

4. If $(1 - \zeta_p)\mathbb{Z}_{\mathbb{Q}(\zeta_p)} \cap \mathbb{Z} \neq p\mathbb{Z}$, deduce that $1 \in (1 - \zeta_p)\mathbb{Z}_{\mathbb{Q}(\zeta_p)}$, so $1 - \zeta_p$ (and all its conjugates) divide 1 in $\mathbb{Z}_{\mathbb{Q}(\zeta_p)}$.

5. Deduce from $(**)$ and the hypothesis in exercise 4 that $1/p$ is in $\mathbb{Z}_{\mathbb{Q}(\zeta_p)}$, and hence in $\mathbb{Z}$, which is a contradiction.

Next we show that, for any $x \in \mathbb{Z}_{\mathbb{Q}(\zeta_p)}$, $\mathrm{Tr}(x(1 - \zeta_p)) \in p\mathbb{Z}$.

6. By factorizing $1 - \zeta_p^i$, show that all the conjugates of $1 - \zeta_p$ are multiples of $1 - \zeta_p$, and hence that the ordinary integer $\mathrm{Tr}(x(1 - \zeta_p))$ is in $p\mathbb{Z}$.

Finally, by writing $x = a_0 + a_1\zeta_p + \cdots + a_{p-2}\zeta_p^{p-2}$ – a combination of the basis elements $1, \zeta_p, \ldots, \zeta_p^{p-2}$ with rational coefficients – and taking the trace of $1 - \zeta_p$ times both sides of this equation, we can show that $a_0, a_1, \ldots$ in turn are in $\mathbb{Z}$, so that integers of $\mathbb{Q}(\zeta_p)$ are in $\mathbb{Z}[\zeta_p]$.

7. Applying Tr to

$$x(1 - \zeta_p) = a_0(1 - \zeta_p) + a_1(\zeta_p - \zeta_p^2) + \cdots + a_{p-2}(\zeta_p^{p-2} - \zeta_p^{p-1}),$$

and observing that $\mathrm{Tr}(\zeta_p^i - \zeta_p^{i+1}) = 0$ by (*) (can you see why?), show that $a_0 p \in p\mathbb{Z}$ and hence $a_0 \in \mathbb{Z}$.
8. Then, since $\zeta_p^{-1} = \zeta_p^{p-1} \in \mathbb{Z}_{\mathbb{Q}(\zeta_p)}$, we have

$$(x - a_0)\zeta_p^{-1} = a_1 + a_2\zeta_p + \cdots + a_{p-2}\zeta_p^{p-3} \in \mathbb{Z}_{\mathbb{Q}(\zeta_p)}.$$

By suitable repetition of the argument in the previous question, show that $a_1, a_2, \ldots \in \mathbb{Z}$, so that $x \in \mathbb{Z}[\zeta_p]$.

## 8.4 Bases and Free Modules

As we saw in Section 6.1, a key feature of an $F$-vector space is the existence of a basis, which allows each $v \in V$ to be expressed uniquely as a linear combination of basis elements with coefficients in the field $F$. In an $R$-module $M$ it may be possible to find **generators** for $M$, in the sense that every member $m \in M$ is a linear combination of generators with coefficients in $R$. But even if the generators are independent, in the sense that no one of them is a linear combination of the others, the expression for $m \in M$ is not necessarily unique.

Examples are the finite Abelian groups, mentioned in Section 8.1 as examples of $\mathbb{Z}$-modules. Specifically, and using the additive notation from Section 3.7, consider the $\mathbb{Z}$-module

$$\mathbb{Z}_2 \oplus \mathbb{Z}_3 = \{a\mathbf{i} + b\mathbf{j} : a, b \in \mathbb{Z}\}$$

$$\text{where} \quad \mathbf{i} = \text{congruence class of 1, mod 2,}$$

$$\mathbf{j} = \text{congruence class of 1, mod 3.}$$

These generators $\mathbf{i}$ and $\mathbf{j}$ are subject to the **relations** $2\mathbf{i} = 0$ and $3\mathbf{j} = 0$. Although $\mathbf{i}$ and $\mathbf{j}$ are independent, their coefficients $a, b$ are not unique, since

$(a + 2)i = ai$ and $(b + 3)i = bi$. Indeed, there are only six different elements $ai + bj$, which is why $\mathbb{Z}_2 \oplus \mathbb{Z}_3$ is a finite Abelian group.

If an $M$-module has generators $m_i$ *not* subject to any relations other than those that hold in any module, such as $m_i + m_j = m_j + m_i$, then $M$ is called a **free** $R$-module. Thus, the $\mathbb{Z}$-module $\mathbb{Z}_2 \oplus \mathbb{Z}_3$ is not free because its generators $i$ and $j$ are subject to relations and, equivalently, their coefficients $a, b \in \mathbb{Z}$ are not unique.

**Definitions.** An $R$-module $M$ is called **free** if there are generators $m_i$ of $M$ such that each $m \in M$ is a linear combination of (finitely many) $m_i$ with unique coefficients $a_i \in R$. A generating set with this property is called a **basis** of $M$, and its size is called the **rank** of $M$ over $R$.

Examples of free $R$-modules are of course the $F$-vector spaces, but also the direct sums $\mathbb{Z} \oplus \mathbb{Z}$, $\mathbb{Z} \oplus \mathbb{Z} \oplus \mathbb{Z}$, and so on, which are free $\mathbb{Z}$-modules, and the ring $\mathbb{Z}_E$ for each field extension $E \supseteq \mathbb{Q}$ of degree $m$, which is a free $\mathbb{Z}$-module of rank $m$, as we saw in the previous section. More generally, for any ring $R$, we define $R^{(n)}$ to be the set of ordered $n$-tuples $(a_1, \ldots, a_n)$, for $a_1, \ldots, a_n \in R$, under the addition operation defined by

$$(a_1, \ldots, a_n) + (b_1, \ldots, b_n) = (a_1 + b_1, \ldots, a_n + b_n),$$

and the scalar multiplication operation defined, for any $c \in R$, by

$$c(a_1, \ldots, a_n) = (ca_1, \ldots, ca_n).$$

Then $R^{(n)}$ is a free $R$-module with $n$ basis elements of the form $(0, \ldots, 1, \ldots, 0)$ (one place equal to 1 and the rest 0). Conversely, any free $R$-module $M$ with a basis $m_1, \ldots, m_n$ is isomorphic to $R^{(n)}$, because each $m \in M$ corresponds to the unique $n$-tuple $(a_1, \ldots, a_n) \in R^{(n)}$ such that $m = a_1 m_1 + \cdots + a_n m_n$.

Free $R$-modules share with vector spaces a notion of dimension, called *rank* in the definition above, because of a theorem: *If $R^{(m)}$ is isomorphic to $R^{(n)}$, then $m = n$.* We proved this theorem in the special case where $R = \mathbb{Z}$ in the previous section. The case of general $R$ requires a rather specialized theorem about determinants, and we do not need it, so we omit the proof.

### 8.4.1 Free $\mathbb{Z}$-modules

Another property that free $R$-modules share with vector spaces is a "subspace" property like that found in Section 6.2. However, we now have to restrict $R$ to be a **principal ideal domain**, mentioned in Section 5.2. In fact, we need the result only for the domain $\mathbb{Z}$, so we will write the proof in terms on $\mathbb{Z}$. (However, the proof is identical with any PID in place of $\mathbb{Z}$.)

**Submodules of a free $\mathbb{Z}$-module.** *Any $\mathbb{Z}$-submodule $M$ of $\mathbb{Z}^{(n)}$ is a free $\mathbb{Z}$-module of rank $m \leq n$.*

*Proof.* The result is trivial for $n = 0$ because $\mathbb{Z}^{(0)} = \{0\}$. Now take $n > 0$, let $u_1, u_2, \ldots, u_n$ be a basis of $\mathbb{Z}^{(n)}$, and suppose the theorem is true for $n - 1$. Then it holds in particular for $\mathbb{Z}^{(n-1)}$, the submodule of $\mathbb{Z}^{(n)}$ generated by $u_2, \ldots, u_n$. So the theorem holds for a submodule $M \subseteq \mathbb{Z}^{(n-1)}$.

Assume now that $M \not\subseteq \mathbb{Z}^{(n-1)}$. The part $M \cap \mathbb{Z}^{(n-1)}$ *is* a submodule of $\mathbb{Z}^{(n-1)}$, so it has a basis $v_2, \ldots, v_m$ with $m - 1 \leq n - 1$ elements. We now seek an element $v_1$ outside $M \cap \mathbb{Z}^{(n-1)}$ that completes a basis for $M$.

Consider the set $I \subseteq \mathbb{Z}$ of "$u_1$ components" of members of $M$:

$$I = \{x \in D : xu_1 + y \in M \text{ for some } y \in \mathbb{Z}^{(n-1)}\}.$$

Then $I$ is a nonzero ideal of $\mathbb{Z}$ and hence, since $\mathbb{Z}$ is a PID, $I$ is generated by some nonzero $d \in \mathbb{Z}$. That is, $I = (d)$. This means there is an element

$$v_1 = du_1 + y_1 \in M \quad \text{for some } y_1 \in \mathbb{Z}^{(n-1)}. \tag{*}$$

To show that $v_1, v_2, \ldots, v_m$ is a basis for $M$, we show:

1. $v_1, v_2, \ldots, v_m$ generate $M$. Well,

$$z \in M \Rightarrow z = xu_1 + y \qquad\qquad \text{for some } x \in I \text{ and } y \in \mathbb{Z}^{(n-1)}.$$
$$\Rightarrow z = d_1du_1 + y \qquad\qquad \text{for some } d_1 \in \mathbb{Z}, \text{ since } I = (d).$$
$$\Rightarrow z - d_1v_1 = (d_1du_1 + y) - d_1(du_1 + y_1) \qquad \text{by definition (*)}.$$
$$\Rightarrow z - d_1v_1 = y - d_1y_1$$
$$\Rightarrow z - d_1v_1 \in M \cap \mathbb{Z}^{(n-1)} \quad \text{since } y, y_1 \in \mathbb{Z}^{(n-1)} \text{ and } z - d_1v_1 \in M.$$
$$\Rightarrow z - d_1v_1 = \sum_{j=2}^{m} d_jv_j \qquad \text{since } v_2, \ldots, v_m \text{ generate } M \cap \mathbb{Z}^{(n-1)}.$$
$$\Rightarrow z = \sum_{j=1}^{m} d_jv_j, \qquad\qquad \text{so } v_1, v_2, \ldots, v_m \text{ generate } M.$$

2. $v_1, v_2, \ldots, v_m$ are independent.
   Suppose that $0 = \sum_{j=1}^{m} d_jv_j$, which gives, by (*),

$$0 = d_1du_1 + d_1y_1 + \sum_{j=2}^{m} d_jv_j.$$

Since $y_1, v_2, \ldots, v_m \in \mathbb{Z}^{(n-1)}$ and $u_1 \notin \mathbb{Z}^{(n-1)}$, it follows that $d_1 d = 0$, hence $d_1 = 0$. But this leaves

$$0 = \sum_{j=2}^{m} d_j v_j.$$

And $v_2, \ldots, v_m$ is a basis, for $M \cap \mathbb{Z}^{(n-1)}$, so $d_2 = \cdots = d_m = 0$ also.   $\square$

Now it is time to reiterate that the free $\mathbb{Z}$-modules we have in mind are the algebraic integer rings $\mathbb{Z}_E$ in field extensions $E$ of finite degree over $\mathbb{Q}$. We have been interested in the ideals of such number rings since Section 5.2, where we found an ideal in the ring $\mathbb{Z}[\sqrt{-5}]$ with the two generators 2 and $1 + \sqrt{-5}$. The theorem above now tells us that *any* ideal of an algebraic integer ring $\mathbb{Z}_E$ is finitely generated, because *an ideal of $\mathbb{Z}_E$ is a $\mathbb{Z}$-submodule*. In particular, this tells us that each ring $\mathbb{Z}_E$ is **Noetherian**, as defined in Section 5.4.

## Exercises

Any $R$-module $M$ can be viewed as the result of imposing relations on a free $R$-module $R^{(m)}$, as we did with the example of $\mathbb{Z}_2 \oplus \mathbb{Z}_3$ above. The idea of "imposing relations" on $R^{(m)}$ can be viewed as forming $M$ as the quotient of $R^{(m)}$ by the $R$-submodule whose elements equal 0 in $M$. Here is how we arrive at $\mathbb{Z}_2 \oplus \mathbb{Z}_3$ from the free $\mathbb{Z}$-module

$$\mathbb{Z}^{(2)} = \{ai + bj : a, b \in \mathbb{Z}\}.$$

1. Let $G = \{2ai + 3bj : a, b \in \mathbb{Z}\}$. Show that $G$ is the free $\mathbb{Z}$-submodule of $\mathbb{Z}^{(2)}$ generated by $2i$ and $3j$.
2. Show that $\mathbb{Z}^{(2)}/G$ is a $\mathbb{Z}$-module in which $2[i] = 3[j] = 0$, where $[i]$ and $[j]$ are the congruence classes, mod $G$, of $i$ and $j$, respectively.
3. Deduce that $\mathbb{Z}^{(2)}/G \cong \mathbb{Z}_2 \oplus \mathbb{Z}_3$.

A nonzero ideal of $\mathbb{Z}_E$ is not just finitely generated, but of the same rank.

4. Let $\omega_1, \ldots, \omega_m$ be an integral basis of $\mathbb{Z}_E$, as found in the previous section. If $I$ is a nonzero ideal of $\mathbb{Z}_E$ and $a \in I$, explain why $a\omega_1, \ldots, a\omega_n \in I$.
5. Conclude that $I$ has the same rank, $m$, as $\mathbb{Z}_E$.

## 8.5 Integers over a Ring

The Noetherian property is one characteristic of algebraic integer rings likely to be useful in studying ideals. Another is the property of "integer" itself, which we now generalize to an arbitrary ring $R$. The obvious way to generalize the concept of integer is to take the algebraic integer concept, from Section 4.6, and replace the (ordinary) integers in its definition by elements of a ring $R$.

**Definition.** If $R$ is a ring and $\alpha$ belongs to some extension of $R$, then $\alpha$ is called **integral over** $R$ if $\alpha$ satisfies an equation of the monic polynomial form

$$x^n + a_{n-1}x^{n-1} + \cdots + a_1 x + a_0 = 0, \quad \text{where } a_0, a_1, \ldots, a_{n-1} \in R.$$

Thus, what we called the *algebraic integers* in Section 4.6 are the integers over $\mathbb{Z}$. As with algebraic integers, it is usually convenient to confine integers over $R$ to a field of finite degree over $R$. To extend $R$ to a field at all, $R$ must be a domain, in which case $R$ embeds in its field of fractions $\mathrm{Frac}(R)$.

**Definition and notation.** If $R$ is a domain and $E \supseteq \mathrm{Frac}(R)$ is an extension field of finite degree, then **integers of** $E$ **over** $R$ are the $\alpha \in E$ that are integral over $R$.

When it is clear which $R$ the integers of $E$ are over, we will simply call them "integers of $E$." These integers form a ring, as the following theorem shows.

**Ring properties of the integers of an extension field.** *If $R$ is a domain and $E \supseteq \mathrm{Frac}(R)$ is an extension field of finite degree, then the integers of $E$ form a ring.*

*Proof.* As in the special case where $R = \mathbb{Z}$ (Section 4.6), it suffices to prove that $\alpha + \beta$, $\alpha - \beta$ and $\alpha\beta$ are integers of $E$ when $\alpha$ and $\beta$ are. (Because then the ring properties of the integers of $E$ are inherited from the ring properties of $E$.)

The proof that $\alpha + \beta$, $\alpha - \beta$ and $\alpha\beta$ are integers is also much the same as when $R = \mathbb{Z}$. This is because the required properties of determinants hold with an arbitrary domain in place of $\mathbb{Z}$, as we remarked in Section 7.3. □

It is interesting and useful to know that all elements of $E$ are actually *fractions* formed from its integers. In fact they are fractions of the special form (integer of $E$)/(member of $R$). The proof is exactly the same as in the special case $R = \mathbb{Z}$, which we proved in Section 8.2, with $R$ in place of $\mathbb{Z}$. So, even in this general setting, elements of a field are "fractions" whose numerator and denominator are "integers" (since elements of $R$ are trivially integers over $R$).

## Exercises

An important application of the integer concept outside domains of numbers is where $R = \mathbb{C}[x]$, the ring of polynomials with complex coefficients.

1. Explain why $\mathbb{C}[x]$ is a domain, so the field $F = \mathrm{Frac}(\mathbb{C}[x])$ exists.

The elements of $F$ are called **rational functions**. (Indeed, in some of the literature, polynomials are called "integral rational functions.") An extension field $E \supseteq F$ of finite degree over $F$ is called an **algebraic function field**.

2. Deduce that an algebraic function $f$ satisfies an equation of the form

$$a_k f^k + \cdots + a_1 f + a_0 = 0, \quad \text{where } a_0, \dots, a_k \in \mathbb{C}[x].$$

3. Explain how each algebraic function $f$ may be realized as a congruence class of polynomials in $f$, illustrating the construction with $f(x) = \sqrt{x}$.

As one would expect, the functions satisfying *monic* polynomial equations with coefficients in $\mathbb{C}[x]$ are called **integral algebraic functions**.

4. Show that $f(x) = 1 + 2\sqrt{x} + x$ is an integral algebraic function of degree 2 over $\mathbb{C}[x]$.
5. Express $g(x) = \frac{1+\sqrt{x}}{1-\sqrt{x}}$ as the quotient of an integral algebraic function by a polynomial, and explain why this is possible for any algebraic function.

## 8.6 Integral Closure

In the previous section we established a convenient setting for studying generalized integers: a domain $R$, an extension field $E \supseteq \mathrm{Frac}(R)$ of finite degree, and the ring of elements of $E$ that are integral over $R$.

**Definition.** Given a domain $R$ and a finite-degree extension $E \supseteq \mathrm{Frac}(R)$, the ring of elements of $E$ that are integral over $R$ is called the **integral closure of** $R$ in $E$ and is denoted by $\mathbb{Z}_{E/R}$.

The word "closure" suggests that $\mathbb{Z}_{E/R}$ is "closed" in some sense, and the sense is given by the following definition.

**Definition.** A domain $D$ is called **integrally closed** if it equals its own integral closure in $\mathrm{Frac}(D)$.

The domain $\mathbb{Z}_{E/R}$ is indeed integrally closed. To prove this, we have to show that if an $x$ in $E$ is integral over $\mathbb{Z}_{E/R}$, then $x$ is already integral over $R$. The proof goes more smoothly if we first establish some equivalent ways of expressing the relation "integral over."

**Equivalents of being integral over $R$.** *If a ring $S \supseteq$ ring $R$ and $x \in S$, then the following properties of $x$ are equivalent:*

*1. x is integral over R,*

*2. R[x] is a finitely generated R-module,*

*3. S has a subring Q that is a finitely generated R-module $\supseteq R[x]$.*

*Proof.* We show that $1 \Rightarrow 2 \Rightarrow 3 \Rightarrow 1$.

$1 \Rightarrow 2$ because an integer $x$ over $R$ satisfies an equation

$$x^n + a_{n-1}x^{n-1} + \cdots + a_1 x + a_0 = 0, \quad \text{where } a_0, \ldots, a_{m-1} \in R.$$

This implies (as previously observed in Section 4.6) that

$$x^n = -a_{n-1}x^{n-1} - \cdots - a_1 x - a_0,$$

so $x^n$ is in the $R$-module generated by $1, x, \ldots, x^{n-1}$. In turn,

$$x^{n+1} = -a_{n-1}x^n - \cdots - a_1 x^2 - a_0 x,$$

so $x^{n+1}$ is also in the $R$-module generated by $1, x, \ldots, x^{n-1}$, since $x^n$ is. Similarly, for all other powers $x^{n+2}, x^{n+3}, \ldots$, so in fact $R[x]$ is the $R$-module generated by $1, x, \ldots, x^{n-1}$.

$2 \Rightarrow 3$ by taking $Q$ to be the $R$-module generated by $1, x, \ldots, x^{n-1}$.

Finally, we show that $3 \Rightarrow 1$ by supposing that there is a finitely generated $R$-module $Q \supseteq R[x]$. Let $v_1, \ldots, v_m$ be generators of $Q$. Then, since $x \in Q$, each $xv_i$ is a linear combination of $v_1, \ldots, v_m$ with coefficients in $R$, so we have the following homogeneous equations in the $v_i$:

$$xv_i = \sum_{j=1}^{m} a_{ij} v_i, \quad \text{where the } a_{ij} \in R.$$

Since these equations have a nonzero solution, their determinant

$$\det \begin{pmatrix} a_{11} - x & a_{12} & \cdots & a_{1m} \\ a_{21} & a_{22} - x & \cdots & a_{2m} \\ \vdots & & & \vdots \\ a_{m1} & a_{m2} & \cdots & a_{mm} - x \end{pmatrix} = 0.$$

Since the highest power of $x$ in the determinant is $\pm x^m$, this gives a monic polynomial equation for $x$, so $x$ is integral over $R$.  □

Next, it is natural to extend the relation "integral over" from elements to whole rings.

**Definition.** If $R$ is a subring of a ring $S$, we call $S$ **integral over** $R$ if each $x$ in $S$ is integral over $R$.

Then $\mathbb{Z}_{E/R}$ is integrally closed by the following proposition.

**Transitivity of "integral over."** *If $T \supseteq S \supseteq R$ are rings such that $T$ is integral over $S$ and $S$ is integral over $R$, then $T$ is integral over $R$.*

*Proof.* If $x \in T$ we have a polynomial equation

$$x^m + b_{m-1}x^{m-1} + \cdots + b_1 x + b_0 = 0, \quad \text{where } b_0, \ldots, b_{m-1} \in S,$$

because $x$ is integral over $S$. The polynomial has coefficients in $R[b_0, \ldots, b_{m-1}]$, so $x$ is integral over $S' = R[b_0, \ldots, b_{m-1}]$.

It follows that $Q = S'[x] = R[b_0, \ldots, b_{m-1}, x]$ is a finitely generated $R$-module containing $R[x]$.

Then $x$ is integral over $R$ by the theorem above and, since $x$ is an arbitrary member of $T$, $T$ is integral over $R$.                                    □

**Corollary.** $\mathbb{Z}_{E/R}$ *is integrally closed.*

*Proof.* $\mathbb{Z}_{E/R}$ is integral over $R$ by definition. Now consider some $x$ in $E$ that is integral over $\mathbb{Z}_{E/R}$. It follows from the proof of transitivity that $x$ is integral over $R$. Hence $x \in \mathbb{Z}_{E/R}$, by definition of $\mathbb{Z}_{E/R}$.                                    □

This corollary applies in particular to the rings $\mathbb{Z}_E$ of integers in algebraic number fields. Thus, we have now established that $\mathbb{Z}_E$ is *integrally closed* as well as being *Noetherian*, which we established in Section 8.4. It remains now to establish a property of the **prime ideals** of $\mathbb{Z}_E$, which we study in the next chapter.

## Exercises

1. Revisit exercise 1 of Section 4.6 to give a direct proof that $\mathbb{Z}$ is integrally closed.

## 8.7 Discussion

As we have seen in this chapter, the initial motivation for studying modules was the theory of algebraic integers and their applications to arithmetic. Dedekind (1871) introduced this theory and rapidly covered everything from the definition of algebraic integer to finding an integral basis for the ring $\mathbb{Z}_E$ of an algebraic number field of degree $n$, thereby showing that $\mathbb{Z}_E$ is (what we would call) a free $\mathbb{Z}$-module of rank $n$. In 1877, disappointed by the lack of response to his ideas, he gave an account for a more general audience, which is now available in English as Dedekind (1996).

In this book, Dedekind began with the concept of module (with the example of algebraic integers in mind) and proceeded to study congruence modulo a submodule and its congruence classes, hence the quotient of modules. Then he specialized to finitely generated modules and their bases, before introducing the concepts of algebraic integers and ideals. Algebraic integers are situated in algebraic number fields of finite degree, and the discriminant is introduced in order to find integral bases for the integers of a field. In particular, finding an integral basis by minimizing the discriminant appears in (Dedekind, 1996, p. 116).

### 8.7.1 Integers over a Domain

The generalization of Dedekind's theory to integers over a domain $R$ was made by Noether (1926), and brought to a wide audience in Chapter 14 of van der Waerden (1931), the pioneering "modern algebra" text. Noether began by defining an element $x$ to be integral over $R$ if its powers belong to an $R$-module of finite rank (number 3 of the three equivalents of "integral" in Section 8.6), evidently finding this more natural than the definition via monic polynomials.

She then proceeded to show her definition equivalent to the "usual conception" of integer, defined by the monic polynomial condition, which had prevailed rather awkwardly since the mid-19th century.

Noether's theory of integers over a domain includes the concept of **integral closure**, which is the second leg of "an abstract characterization of those rings whose ideal theory coincides with that of the integers of an algebraic number

Figure 8.1 Bartel Leendert van der Waerden (1903–1996) (public domain).

Figure 8.2  Heinrich Martin Weber (1842–1913) (public domain).

field." The first leg was the Noetherian property of Section 5.4, which she introduced in Noether (1921). Thus Noether is looking for the "right concepts" to capture unique prime ideal factorization. To complete the list of concepts, one needs a condition on the ideals, which we study in the next chapter. That condition, too, is motivated by the properties of algebraic number rings.

### 8.7.2  Rings of Algebraic Functions

Noether (1926) also remarked that her theory applies to integers over **algebraic function fields** ("integral algebraic functions"). We have sketched the basic ideas of these fields in the exercises to Section 8.5, and will say more about them in Section 9.8.

The extension of Dedekind's theory of algebraic integers to function fields was made by Dedekind and Weber (1882), in a paper now available in English translation as Dedekind and Weber (2012). As Noether realized, this extension reveals the potentially vast applicability of ideal theory to algebraic geometry.

# 9

# Ideals and Prime Factorization

## Preview

In previous chapters we have extended the concepts of addition, multiplication, rational numbers, and integers to the general setting of rings and fields. This setting brings to light the concepts of vector space and module, which clarify the nature of many rings and fields by giving them a somewhat geometric structure—in particular a concept of "dimension."

With this structural understanding of rings, we can now return to the problem that provoked all these developments: how to define a notion of "prime" for which "unique prime factorization" is true. We hope to have this notion at least for the rings $\mathbb{Z}_E$ of algebraic integers, which serve as a microscope for studying the ordinary integers.

The appropriate concept of "prime" is that of a **prime ideal**, and the rings admitting "unique factorization" of prime ideals are called **Dedekind domains**. We already have all the concepts needed to define Dedekind domains and to prove their basic properties, except for the concept of prime ideal itself. Defining prime ideals is actually easy to do: it is motivated by Euclid's result that if a prime $p$ divides a product $ab$, then $p$ divides $a$ or $p$ divides $b$.

The concept of Dedekind domain appears complicated at first, involving the concepts of integrality, finite-dimensionality, and prime ideals. But in proving the existence of unique prime factorization, we invoke another concept – that of **inverse ideals** – which turns out to characterize the Dedekind property exactly and concisely.

In a reasonable sense, this characterization shows that Dedekind domains are the "right concept" to explain unique prime factorization.

## 9.1 To Divide Is to Contain

The simplest ideals that we have seen (in Section 5.2) are the *principal ideals*, which indeed are the only ideals in certain rings. These are the **principal ideal domains** (PIDs), and the first example is $\mathbb{Z}$. A principal ideal is one of the form

$$(a) = \{ra : r \in R\},$$

and it closely imitates the element $a$ itself, as far as divisibility is concerned. In particular

$$b \text{ divides } a \Leftrightarrow a = bc \quad \text{for some } c \in R,$$
$$\Leftrightarrow ra = (rc)b \in (b) \quad \text{for any } r \in R,$$
$$\Leftrightarrow (b) \supseteq (a).$$

So, for principle ideals, we can say that "to divide is to contain."

For ideals in arbitrary rings, containment is the *definition* of "divides."

**Definition.** If $I$ and $J$ are ideals of a ring $R$, then we say that $I$ **divides** $J$ if $I \supseteq J$.

With this definition of "divides," what should the greatest common divisor be? Well, first, note that $R$ is itself an ideal that contains all ideals, so $R$ *divides* all ideals, and so it should be the unit ideal. Indeed, $R$ is the principal ideal $(1)$. More generally, we have a gcd of any two ideals.

**Greatest common divisor of two ideals.** *If $I$ and $J$ are ideals of a ring $R$, then* $\gcd(I, J) = \{ri + sj : i \in I, j \in J, \text{ and } r, s \in R\}$.

*Proof.* By definition of divisibility, any ideal $L$ that divides (that is, contains) both $I$ and $J$ should divide (that is, contain) their gcd. Thus $\gcd(I, J)$ should be the "smallest" ideal containing $I$ and $J$, which means that $\gcd(I, J)$ is the intersection of all ideals containing both $I$ and $J$.

This set is

$$K = \{ri + sj : i \in I, j \in J; r, s \in R\}. \tag{*}$$

It is clear that $K$ is closed under sums, and under products by members of $r$, so $K$ is an ideal. Also, $K$ is the "least" such ideal, because any ideal containing $I$ and $J$ must contain the sums of all their members, and such sums are precisely the members of $K$. □

In the PID $\mathbb{Z}$, this theorem says that

$$\gcd((a), (b)) = \{ma + nb : m, n \in \mathbb{Z}\}.$$

And, as we know from Section 1.2,

$$\{ma + nb: m, n \in \mathbb{Z}\} = \{\text{multiples of } \gcd(a, b)\} = (\gcd(a, b)).$$

We have seen a more interesting example in $\mathbb{Z}[\sqrt{-5}]$, where

$$I = \left\{2m + (1 + \sqrt{-5})n: m, n \in \mathbb{Z}\right\}$$
$$= \left\{2\mu + (1 + \sqrt{-5})v: \mu, v \in \mathbb{Z}[\sqrt{-5}]\right\}$$

is a *non*principal ideal. In Section 5.1, we found this ideal by looking for the gcd of 2 and $1 + \sqrt{-5}$ in $\mathbb{Z}[\sqrt{-5}]$, and indeed it is the gcd of the principal ideals $(2)$ and $(1 + \sqrt{-5})$ by formula (*).

**Remark.** Now that "divides" is defined to mean "contains," the ascending chain condition (ACC) defining Noetherian rings in Section 5.4 says that there is no infinite *descending* chain of divisors. Indeed, Noether (1926) called ACC the "divisor chain theorem." Finiteness of divisor chains is certainly a natural assumption when one hopes to prove the existence of prime factorization. It is a distant echo of the argument used by Euclid (Section 1.1) to prove the existence of prime factorization in $\mathbb{N}$.

Next, we have to define "prime" for ideals.

### Exercises

1. Show that the ideal $J$ in the exercises to Section 5.1 equals the gcd of $(3)$ and $(1 + \sqrt{-5})$.

$I$ and $J$ also have the **conjugate ideals**:

$$\overline{I} = \left\{2m + (1 - \sqrt{-5})n: m, n \in \mathbb{Z}\right\},$$
$$\overline{J} = \left\{3m + (1 - \sqrt{-5})n: m, n \in \mathbb{Z}\right\}$$

2. Check that $\overline{I}$ and $\overline{J}$ are ideals by verifying their closure under multiplication by $\mu \in \mathbb{Z}[\sqrt{-5}]$.
3. Of which principal ideals is $\overline{I}$ (and similarly $\overline{J}$) the gcd?
4. Since "to divide is to contain," how should the **least common multiple** (lcm) of two ideals be defined?
5. Test your definition of lcm on the principal ideals $(a)$ and $(b)$ in $\mathbb{Z}$.

## 9.2 Prime Ideals

Before stating the definition of a prime ideal in an arbitrary ring, let us recall Euclid's result about prime divisors $p \in \mathbb{Z}$ (Section 1.3).

**Prime divisor property.** *If prime $p$ divides $ab$, then $p$ divides $a$ or $p$ divides $b$.*                                                                                    □

An equivalent statement in terms of ideals in $\mathbb{Z}$ is:

$$\text{If } ab \in (p), \text{ then } a \in (p) \text{ or } b \in (p).$$

This equivalent of the prime divisor property prompts the definition of **prime ideal**:

**Definition.** An ideal $I$ in a ring $R$ is **prime** if, for any $a, b \in R$, $ab \in I$ implies $a \in I$ or $b \in I$.

Thus, the principal ideal $(p)$ is a prime ideal of $\mathbb{Z}$ by the prime divisor property. An example of a prime ideal which is not principal is the ideal of $\mathbb{Z}[\sqrt{-5}]$,

$$I = \left\{ 2m + (1 + \sqrt{-5})n : m, n \in \mathbb{Z} \right\}.$$

In Section 5.4 we found that $I$ is a *maximal* ideal, and this implies it is prime, because *all maximal ideals are prime*. This fact can be proved directly, but a more enlightening proof deduces it from a characterization of prime ideals like the characterization of maximal ideals in Section 5.4.

**Characterization of prime ideals.** *An ideal $I$ in a ring $R$ is prime if and only if $R/I$ is a domain.*

*Proof.* Observe the following equivalences, where $[x]$ denotes the congruence class of $x \in R$, mod $I$:

$$I \text{ is a prime ideal of } R \Leftrightarrow (ab \in I \Rightarrow a \in I \text{ or } b \in I)$$
$$\Leftrightarrow ([a][b] = [0] \Rightarrow [a] = [0] \text{ or } [b] = [0])$$
$$\Leftrightarrow R/I \text{ has no zero divisors.}$$
$$\Leftrightarrow R/I \text{ is a domain.} \qquad \square$$

**Corollary.** *A maximal ideal is prime.*

*Proof.* If $I$ is a maximal ideal of $R$, then $R/I$ is a field, by the characterization of maximal ideals in Section 5.4. And a field is a domain.                              □

One might think that maximality ought to be the same as prime. After all, since "to contain is to divide," a maximal ideal is one divisible only by itself and the whole ring, which is the principal ideal $(1)$. However, the prime divisor property is what we really demand of primes, and this is a weaker property than maximality.

Not every prime ideal is maximal, as we can show by finding an example where $R/I$ is a domain but not a field. This happens in the polynomial ring $R = \mathbb{Z}[x]$, where $I = (x)$ is an ideal and $R/I \cong \mathbb{Z}$ (since quotienting by $(x)$ amounts to setting $x = 0$). However, in the case that motivates us – rings $\mathbb{Z}_E$ of algebraic integers – nonzero prime ideals are maximal, as we will see in Section 9.4.

Thus, we are motivated to study rings in which all prime ideals are maximal, as well as having the other key properties of rings of algebraic integers: being integrally closed and Noetherian. These rings were first singled out by Noether (1926), and these particular defining conditions were given by Krull (1928).

**Definition.** A **Dedekind domain**[1] is a domain that is integrally closed, Noetherian, and in which all nonzero prime ideals are maximal.

## Exercises

A direct proof that maximal ideals are prime is quite short, and it involves an argument similar to the proof of the prime divisor property in Section 1.3.

1. Suppose $I$ is a maximal ideal in $R$, $ab \in I$, and $a \notin I$. Show that the ideal $\{ma + i : m \in \mathbb{Z}, i \in I\} = R$ and, in particular, $1 = ma + i$ for some $m \in \mathbb{Z}$ and $i \in I$.

2. Deduce from exercise 1 that $b \in I$, and hence that $I$ is a prime ideal.

$\mathbb{Z}[x]$ is also a simple example of a ring that is not a PID (unlike $\mathbb{Q}[x]$, or $F[x]$ for any field $F$).

3. Suppose on the contrary that the ideal $\{2m + xn : m, n \in \mathbb{Z}[x]\}$ of $\mathbb{Z}[x]$ is a principal ideal $(p)$. Deduce that $p$ divides $2$ and $p$ divides $x$, and derive a contradiction.

---

[1] These are also known as **Dedekind rings**, but it seems useful to be reminded that being a domain is part of their definition.

## 9.3  Products of Ideals

We were able to define prime ideals, rather surprisingly, without defining the **product** of ideals. However, we will need this product, and its definition is:

**Definition.** The **product of ideals** $A$ and $B$ in a ring $R$ is the set of finite sums of products $ab$, where $a \in A$ and $b \in B$. That is

$$AB = \{a_1 b_1 + \cdots + a_n b_n : a_1, \ldots, a_n \in A; b_1, \ldots, b_n \in B\}.$$

With this definition, we can restate the prime divisor property for ideals:

**Prime ideal divisor property.** *If a prime ideal $P$ divides a product $AB$ of ideals, then $P$ divides $A$ or $P$ divides $B$.*

*Proof.* Suppose that $P \supseteq AB$ but that $P \not\supseteq A$. We wish to prove that $P \supseteq B$.

Since $P \not\supseteq A$, there is an $a \in A$ with $a \notin P$. For any $b \in B$ we have $ab \in AB \subseteq P$, so $b \in P$ by definition of prime ideal. This means $P \supseteq B$, as required.                                                                                           □

We can in fact take the prime ideal divisor property as the *definition* of prime ideal. If $I$ has the prime ideal divisor property, and if $ab \in I$ and $a \notin I$, then $b \in I$ by considering the principal ideals $A = (a)$ and $B = (b)$.

It might now seem that we are close to proving unique prime factorization for ideals. But, not so fast. The trouble is that *existence* of prime factorization is no longer obvious. Existence depends on properties of Dedekind domains, starting with the Noetherian property.

**Prime ideal products in Noetherian domains.** *In a Noetherian domain $R$ each nonzero ideal contains a product of prime ideals.*

*Proof.* If there are nonzero ideals of $R$ *not* containing products of prime ideals, then there is a *maximal* one among them, $B$, since $R$ is Noetherian.

$B$ is not prime, by hypothesis, so there are $x, y \notin B$ with $xy \in B$. Then the ideals $B + Rx$ (the sums of elements from $B$ and $Rx$) and $B + Ry$ (similarly) properly contain $B$ so, by the maximal property of $B$,

$$B + Rx \supseteq P_1 \cdots P_m, \quad B + Ry \supseteq Q_1 \cdots Q_n,$$
$$\text{for prime ideals } P_1, \ldots, P_m, Q_1, \ldots, Q_n.$$

But now, since $xy \in B$, we have $Rxy \subseteq B$. It follows that $(B+Rx)(B+Ry) \subseteq B$, and therefore

$$B \supseteq (B + Rx)(B + Ry) \supseteq P_1 \cdots P_m \cdot Q_1 \cdots Q_n,$$

contrary to the definition of $B$. So each nonzero ideal of $R$ contains a product of prime ideals. □

We note that, since "to divide is to contain," the conclusion of the theorem says that *each nonzero ideal $I$ divides a product $P_1 \cdots P_m$ of prime ideals*. One is tempted to conclude that

$$P_1 \cdots P_m = IJ \quad \text{for some ideal } J.$$

However, we have not yet proved that a "divisor" $I$ of some $K$ is necessarily a *factor of a product $IJ$ equal to $K$*. This will emerge indirectly in Section 9.6, after we introduce "inverse" ideals in Section 9.5.

## Exercises

With the definition of product of ideals we are finally able to explain why the different factorizations of 6 in $\mathbb{Z}[\sqrt{-5}]$, namely $2 \cdot 3$ and $(1+\sqrt{-5})(1-\sqrt{-5})$, can be split into the *same* product of prime ideals. We have already established that the ideals $I$ and $J$ from Section 5.1 are maximal, and hence prime.

1. Show similarly that the ideals $\overline{I}$ and $\overline{J}$ are prime.

We have also established that

$$I = \gcd\left((2), (1+\sqrt{-5})\right),$$
$$J = \gcd\left((3), (1+\sqrt{-5})\right),$$
$$\overline{I} = \gcd\left((2), (1-\sqrt{-5})\right),$$
$$\overline{J} = \gcd\left((3), (1-\sqrt{-5})\right).$$

2. Show that all members of the ideal $I^2$ are multiples of 2, and that they include 2. Hence, $I^2 = (2)$.
3. Show that all members of the ideal $J\overline{J}$ are multiples of 3, and that they include 3. Hence, $J\overline{J} = (3)$.

We now have the prime ideal factorization $(2) \cdot (3) = I^2 J \overline{J}$, and we can check that this is a prime ideal factorization of $(1 + \sqrt{-5})(1 - \sqrt{-5})$ as follows.

4. Show that the principal ideal $(1 + \sqrt{-5}) = IJ$.
5. Show similarly that $(1 - \sqrt{-5}) = \overline{I}\,\overline{J}$, and finally that $\overline{I} = I$, so

$$(1 + \sqrt{-5})(1 - \sqrt{-5}) = I^2 J \overline{J}.$$

## 9.4 Prime Ideals in Algebraic Number Rings

To show that the theory of Dedekind domains applies to the algebraic integer rings $\mathbb{Z}_E$, it remains to show that the nonzero prime ideals of $\mathbb{Z}_E$ are maximal, since we know that $\mathbb{Z}_E$ is Noetherian by Section 8.4.1 and integrally closed by Section 8.6. To do this, we begin with a basic lemma about domains.

**Finite domains.** *Any finite domain is a field.*

*Proof.* Suppose $D$ is a finite domain with nonzero elements $d_1, \ldots, d_k$. Then if $d$ is any nonzero element of $D$, each $dd_i \neq 0$ because $D$ has no zero divisors. Also, if $d_i \neq d_j$, then $dd_i \neq dd_j$ because $dd_i = dd_j$ implies $d(d_i - d_j) = 0$, which is impossible because $D$ has no zero divisors.

It follows that $dd_1, \ldots, dd_k$ are the nonzero elements of $D$ again, at worst in a different order. In particular, some $dd_i = 1$. Thus, each nonzero element $d \in D$ has an inverse, so $D$ is a field.                    $\square$

**Prime ideals of $\mathbb{Z}_E$.** *If $E \supset \mathbb{Q}$ is a field extension of finite degree, then each nonzero prime ideal in $\mathbb{Z}_E$ is maximal.*

*Proof.* Suppose $P \subseteq \mathbb{Z}_E$ is a prime ideal. Then $\mathbb{Z}_E/P$ is a domain by the characterization of prime ideals in the previous section. It now suffices to prove that the domain $\mathbb{Z}_E/P$ is finite, and hence a field, since this implies $P$ is maximal by the characterization of maximal ideals in Section 5.4.

To do this, we choose a nonzero $x \in P$ and use it to find a nonzero ordinary integer $a \in P$. Since $x \in \mathbb{Z}_E$, we have a minimal equation

$$x^n + a_{n-1}x^{n-1} + \cdots + a_1 x + a_0 = 0, \quad \text{where } a_0, \ldots, a_{n-1} \in \mathbb{Z}.$$

Notice first that $a_0 \neq 0$, otherwise we could divide the equation by $x$, contrary to minimality. Then rearranging gives

$$a_0 = x\left(-a_1 - \cdots - a_{n-1}x^{n-2}\right),$$

which shows that $a_0$ is a multiple of $x$, and hence $a_0 \in P$. Thus, $P$ includes the nonzero ordinary integer $a_0$, which we will call $a$.

Now recall from Section 8.2 that $\mathbb{Z}_E$ has a finite integral basis $\omega_1, \ldots, \omega_m$. Since $a$ is in $P$, its multiples $a\omega_1, \ldots, a\omega_m$ are also in $P$. It follows, by adding or subtracting such multiples, that each element $b_1\omega_1 + \cdots + b_m\omega_m \in \mathbb{Z}_E$ is congruent, mod $P$, to one with $|b_1|, \ldots, |b_m| \leq a$.

There are only finitely many elements of this form, and therefore $\mathbb{Z}_E/P$ is finite, as required.                    $\square$

With this theorem we have proved that the ring of integers $\mathbb{Z}_E$ in any algebraic number field is a Dedekind domain. Thus, the results we now prove about Dedekind domains, notably unique prime ideal factorization, are guaranteed to apply where number theory wants them.

## Exercises

1. Using the fact, from the exercises to Section 8.4, that a nonzero ideal $I$ of $\mathbb{Z}_E$ has the same rank as $\mathbb{Z}_E$, show that $\mathbb{Z}_E/I$ is finite for *any* nonzero $I$.
2. Why is the ring $\mathbb{Z}_E/I$ not necessarily a field?
3. Find how many elements are in $\mathbb{Z}[\sqrt{-5}]/(2)$.
4. Find elements that show $\mathbb{Z}[\sqrt{-5}]/(2)$ is not a field.

## 9.5 Fractional Ideals

Our first step in the general theory of Dedekind domains is to expand the concept of ideal with an eye towards creating *inverse* ideals and legitimizing *cancellation* in products of ideals.

**Definition.** If $R$ is a domain, then an $R$-submodule $I \subseteq \mathrm{Frac}(R)$ is called a **fractional ideal** if there is a nonzero $d \in R$ such that $dI \subseteq R$.

Thus, the elements of $I$ have $d$ as a "common denominator." Ordinary (or "integral") ideals are those with $d = 1$. In any case, since $I$ is an $R$-submodule of $\mathrm{Frac}(R)$, $dI$ is an $R$-submodule of $R$, and hence an ideal. If $R$ is Noetherian, then $dI$ is finitely generated, and hence so is $I$ as an $R$-submodule of $\mathrm{Frac}(R)$.

We use fractional ideals to obtain *inverses* of ideals in a Dedekind domain. An inverse $I'$ of ideal $I$ satisfies $II' = R$, since $R$ is the identity for multiplication of ideals (defining the product of fractional ideals the same way as for ideals.)

**Inverses of maximal ideals.** *In a Dedekind domain $R$ that is not a field, each maximal $M$ ideal has an inverse $M'$.*

*Proof.* Let $M$ be a maximal ideal of $R$. Then $M \neq (0)$ since $R$ is not a field. Let

$$M' = \{x \in \mathrm{Frac}(R) : xM \subseteq R\}. \tag{*}$$

Then $M'$ is an $R$-submodule of $\mathrm{Frac}(R)$. Any $d \in M$ is a common denominator for all $x \in M'$, so $M'$ is a fractional ideal. It remains to show that $MM' = R$.

Certainly, the definition (*) implies $MM' \subseteq R$. Also $R \subseteq M'$ because $M$ is an ideal, hence $xM \subseteq M \subseteq R$ for all $x \in R$. Since $M$ is maximal and

$$M \subseteq MM' \subseteq R$$

we have either $MM' = R$ or $MM' = M$. It now remains to show $MM' \neq M$.

If, on the contrary, $MM' = M$, then

$x \in M' \Rightarrow xM \subseteq M$

$\quad\quad \Rightarrow x^2 M \subseteq M \cdots$

$\quad\quad \Rightarrow x^n M \subseteq M$ for any $n$, by induction

$\quad\quad \Rightarrow$ any nonzero $d \in M$ is a common denominator for all $x^n$

$\quad\quad \Rightarrow R[x]$ is a fractional ideal of $R$.

Since $R$ is Noetherian, the fractional ideal $R[x]$ is a finitely generated $R$-module by the remark above (following the definition of fractional ideal). This means $x$ is *integral* over $R$ by the equivalents of "integral over" proved in Section 8.6. But $R$ is Dedekind, hence integrally closed, so $x \in R$. Thus, $MM' = M$ leads to $M' = R$, which we finally have to show is impossible.

Take a nonzero $a \in M$. The ideal $Ra$ contains a product $P_1 \cdots P_m$ of nonzero prime ideals by the theorem on prime ideal products in Section 9.2. Take $m$ as small as possible and notice that

$$M \supseteq Ra \supseteq P_1 \cdots P_m \Rightarrow M \supseteq \text{ some } P_i$$

by the prime ideal divisor property of Section 9.2. Say $M \supseteq P_1$ by suitable renumbering. If we put $B = P_2 \cdots P_m$, then

$$Ra \supseteq MB \text{ and } Ra \not\supseteq B \text{ since } m \text{ is minimal.}$$

Thus, there is a $b \in B$ with $b \notin Ra$. But

$$MB \subseteq Ra \Rightarrow Mb \subseteq Ra \Rightarrow Mba^{-1} \subseteq R \Rightarrow ba^{-1} \in M',$$

by definition of $M'$. Yet $b \notin Ra$ implies $ba^{-1} \notin R$, so $R \neq M'$, as required.

$\square$

Notice that this proof of the existence of inverses uses the Dedekind domain properties to the full: the Noetherian property, integral closure, and the maximality of prime ideals. In Section 9.7 we will show that these are in fact the "right" properties to prove existence of inverses, because existence of inverse ideals implies the Dedekind domain properties.

**Remark.** The existence of an inverse ideal $I^{-1}$ is *obvious* in the case of a principal ideal $I = Rd$ for some nonzero $d \in R$. In this case it is clear that

$Rd^{-1}$ is a fractional ideal and that $(Rd)(Rd^{-1}) = R$, so $Rd^{-1} = (Rd)^{-1}$. We use inverses of principal ideals in the next section to prove, among other things, that *all nonzero fractional ideals of a Dedekind domain have inverses.*

## Exercises

Knowing the inverses of principal ideals, we can also calculate the inverses of certain other ideals.

1. Describe the inverse ideal $(2)^{-1}$ in both $\mathbb{Q}$ and $\mathbb{Q}(\sqrt{-5})$.
2. Explain why $I^2(2)^{-1} = \mathbb{Z}[\sqrt{-5}]$, where $I = \{2m + (1 + \sqrt{-5}): m,$ $n \in \mathbb{Z}\}$, and hence find $I^{-1}$.
3. Similarly find $J^{-1}$, where $J = \{3m + (1 + \sqrt{-5}): m, n \in \mathbb{Z}\}$.

## 9.6 Prime Ideal Factorization

We now combine the existence of a product of prime ideals divisible by a given ideal (Section 9.2) with inverses of maximal ideals from the previous section to prove *existence and uniqueness of prime ideal factorization*. It follows that the nonzero fractional ideals form an Abelian group.

**Unique prime ideal factorization.** *If $R$ is a Dedekind domain, then each nonzero fractional ideal $I \subseteq Frac(R)$ is uniquely expressible in the form*

$$I = P_1^{n_1} \cdots P_m^{n_m}, \text{ for some prime ideals } P_1, \ldots, P_m \neq (0) \text{ and } n_1, \ldots, n_m \in \mathbb{Z},$$

*and the nonzero fractional ideals form an Abelian group under the product operation. Also, if $I \subseteq R$, the exponents $n_i$ are nonnegative.*

*Proof.* By definition of the fractional ideal $I$, $dI \subseteq R$ for some nonzero $d \in R$. Since $dI$, $Rd$ are integral ideals with $I = (dI)(Rd)^{-1}$, it suffices to prove unique prime factorization of integral ideals. We first prove *existence* of the factorization.

If $R$ contains ideals that are *not* products of prime ideals, there is a *maximal* one among them, $J$, since $R$ is Noetherian. Then $J \neq R$, since $R = MM^{-1}$ for any maximal (and hence prime) ideal $M$ by the theorem in the previous section. So $J \subseteq$ maximal ideal $P$, the maximal ideal of $R$ among those containing $J$.

The maximal $P$ has an inverse $P^{-1}$, and $P$ is prime since $R$ is Dedekind. Also,

$$J \subseteq P \Rightarrow JP^{-1} \subseteq PP^{-1} = R, \text{ and}$$

$$P^{-1} \supseteq R \Rightarrow JP^{-1} \supseteq J, \text{ so we have } J \subseteq JP^{-1} \subseteq R.$$

Now $JP^{-1} \neq J$ because

$$JP^{-1} = J \text{ and } x \in P^{-1} \Rightarrow xJ \subseteq J$$
$$\Rightarrow x^n J \subseteq J \text{ for all } n$$
$$\Rightarrow x \text{ integral over } R$$
$$\Rightarrow x \in R \text{ as in the previous theorem.}$$

And this is impossible, because $P^{-1} \subseteq R \Rightarrow PP^{-1} \subseteq P \neq R$.

Thus we have $J \subset JP^{-1}$, so the maximality of $J$ implies

$$JP^{-1} = P_1 \cdots P_m \text{ for some prime ideals } P_1, \ldots, P_m,$$

and therefore

$$J = PP_1 \cdots P_m, \text{ contradicting the definition of } J.$$

So in fact *each ideal is the product of prime ideals*. Uniqueness follows by the prime ideal divisor property of Section 9.2 (just as it did for $\mathbb{Z}$ in Section 1.3).

It also follows that *each nonzero fractional ideal has an inverse*, obtained by reversing the signs of all the exponents in its prime factorization. Then, since the product of ideals is obviously associative and commutative, the nonzero fractional ideals form an Abelian group.                                    □

Another consequence of the existence of inverses is that if an ideal $J$ divides an ideal $I$, then $I = JK$ for some ideal $K$. In fact, we have:

**Division implies a product.** *If $I$ and $J$ are ideals in $R$ and $J^{-1}$ exists in Frac($R$), then there is an ideal $K \subseteq R$ with $I = JK$ if and only if $J \supseteq I$ ($J$ divides $I$).*

*Proof.* If $J \supseteq I$, the solution $K = IJ^{-1}$ of $I = JK$ is an ideal of $R$ because

$$J \supseteq I \Rightarrow JJ^{-1} \supseteq IJ^{-1} \Rightarrow R \supseteq K.$$

Conversely, if $I = JK$ for an ideal $K \subseteq R$ then $J \supseteq I$ because

$$R \supseteq K \Rightarrow JR \supseteq JK \Rightarrow J \supseteq I.$$                                    □

This **divisibility lemma** finally gives the answer to the question, raised in Section 9.2, whether the concept of "divides" has the right relationship with the concept of "product."

Moreover, the divisibility lemma enables us to prove that prime ideals do not split into nontrivial products of ideals, so the concept of "prime" also has the right relationship with the concept of "product."

**Indivisibility of prime ideals.** *If $P$ is a nonzero prime ideal of a ring $R$ whose nonzero ideals have inverses, then $P = AB$ only if $A = P$ or $B = P$.*

*Proof.* $P = AB$ implies $P \supseteq AB$, in which case $P \supseteq A$ or $P \supseteq B$, by the prime ideal divisor property of Section 9.2.

But $P = AB$ and $A \subseteq R$ implies $B \supseteq P$ by the divisibility lemma. If $P \supseteq B$, this gives $B = P$, and hence $A = R$ by multiplying each side of $P = AB$ by $B^{-1}$. Otherwise $P \supseteq A$, in which case $B \subseteq R$ similarly implies $A = P$ and $B = R$. $\square$

## Exercises

Another way in which divisibility of ideals has the "right" relationship with the concept of product may be seen with the concepts of greatest common divisor and least common multiple.

1. If $A = P_1^{n_1} \cdots P_k^{n_k}$, for prime ideals $P_1, \ldots, P_k \neq (0)$, is an integral ideal with positive $n_1, \ldots, n_k \in \mathbb{Z}$, and if $C$ divides $A$, deduce from the divisibility lemma that

$$C = P_1^{l_1} \cdots P_k^{l_k}, \quad \text{for nonnegative } l_1 \leq n_1, \ldots, l_k \leq n_k \in \mathbb{Z}.$$

2. Deduce from exercise 1 that if $P_1, \ldots, P_k$ are the prime ideals in the prime ideal factorization of an integral ideal $A$ (as above, but with nonnegative exponents $n_i$) and

$$B = P_1^{m_1} \cdots P_k^{m_k}, \quad \text{for nonnegative } m_1, \ldots, m_k \in \mathbb{Z},$$

then

$$\gcd(A, B) = P_1^{\min(m_1, n_1)} \cdots P_k^{\min(m_k, n_k)}.$$

3. Prove the corresponding result for least common multiple (lcm) and hence show

$$\gcd(A, B)\operatorname{lcm}(A, B) = AB.$$

## 9.7 Invertibility and the Dedekind Property

In the previous section we showed that any Dedekind domain $R$ has unique prime ideal factorization and in $\operatorname{Frac}(R)$ there is an inverse for each nonzero fractional ideal. We now show that the latter property *implies* the Dedekind property, so they are actually equivalent, as was discovered by Krull (1935).[2] Invertibility has already given quick answers to questions on the relation between *divisibility* (in the sense of containment) and *products* of ideals. Invertibility also makes short work of the Dedekind properties.

**Invertibility implies the Dedekind properties.** *If R is a ring whose nonzero ideals have inverses, then R is a Dedekind domain; that is,*

1. *it is Noetherian,*
2. *it is integrally closed, and*
3. *each of its nonzero prime ideals is maximal.*

*Proof.* First observe that $R$ must be a domain to have the invertibility property, because if nonzero $a, b \in R$ and $ab = 0$, then the nonzero ideals $(a)$ and $(b)$ have product $(0)$, so neither $(a)$ nor $(b)$ can have an inverse. We now see why invertibility implies the properties 1, 2, and 3.

To prove 1, suppose $I \neq (0)$ is an ideal of $R$, with inverse $I'$. Then since $II' = R$, which includes 1, there must be $x_1, \ldots, x_n \in I$ and $x'_1, \ldots, x'_n \in I'$ with

$$x_1 x'_1 + \cdots + x_n x'_n = 1.$$

Now take any $x \in I$ and notice that this equation gives

$$x \in I \Rightarrow x = (x x'_1 x_1 + \cdots + x x'_n x_n)$$
$$\Rightarrow x \in R x_1 + \cdots + R x_n$$
$$\text{because each } x x'_i \in II' = R$$
$$\Rightarrow x_1, \ldots, x_n \text{ generate } I.$$

Thus, each ideal of $R$ is finitely generated, so $R$ is Noetherian.

To prove 2, suppose that $u$ is integral over $R$ and that $u \in \mathrm{Frac}(R)$. Then, on the one hand, $u$ satisfies an equation

$$x^n + a_{n-1} x^{n-1} + \cdots + a_0 = 0, \quad \text{where } a_0, \ldots, a_{n-1} \in R.$$

So $R + Ru + \cdots + Ru^{n-1}$ includes $u^n$, hence also $u^{n+1}, u^{n+2}, \ldots$, and indeed all of $R[u]$, by the usual argument. On the other hand, $u = a/b$ for some $a, b \in R$, so $b^{n-1}$ is a common denominator of $1, u, \ldots, u^{n-1}$ and hence for all of $R[u]$. It follows that $R[u]$ is a nonzero fractional ideal $I \subseteq \mathrm{Frac}(R)$. It is

---

[2] However, it should be mentioned the Noether (1926) had already shown unique prime ideal factorization equivalent to her formulation of the Dedekind domain properties.

also clear, since $I$ consists of the polynomials in $u$, that $II = I$. Now, if each nonzero fractional ideal has an inverse, multiplying $II = I$ by an inverse of $I$ gives $I = R$.

In particular, since $u \in I$ implies $u \in R$, $R$ is integrally closed.

To prove 3, suppose that $P$ is a nonzero prime ideal of $R$. If $P$ is not maximal, then there is an ideal $Q$ with $P \subset Q \subset R$. This implies, by the divisibility lemma of the previous section, that $P = KQ$ for some ideal $K$. Also, $K \neq R$ because $P \neq Q$. So in the product $P = KQ$, we have $K \neq R$ and $Q \neq R$, contradicting the indivisibility of prime ideals proved at the end of the last section. □

Now that we have proved that invertibility implies the Dedekind domain properties, which imply unique prime ideal factorization, we could close the circle by proving that unique prime ideal factorization implies invertibility. Assuming unique prime ideal factorization of *fractional* ideals, as proved in the previous section, the existence of inverses is obvious. As seen there, we obtain the inverse of a nonzero fractional ideal by reversing the signs of all the exponents in its prime ideal factorization.

If we assume unique prime ideal factorization only for *integral* ideals – that is, ideals $I \subseteq R$ – then it is still possible to show invertibility of nonzero ideals. The proof may be seen in (Zariski and Samuel, 1958, p. 273). However, the proof involves some tricky manipulations, which we take as a sign that the "right" statement of unique prime ideal factorization is the one involving fractional ideals.

## Exercises

Unique prime ideal factorization has become the main focus of this chapter, as we searched for the abstract conditions that make it possible. However, we do not wish to forget the problems that called for prime ideal factorization in the first place. Here is one we can now solve.

First, recall the example in Section 2.9, where Euler worked in $\mathbb{Z}[\sqrt{-2}]$ to find all the (ordinary) integer solutions of

$$y^3 = x^2 + 2 = (x + \sqrt{-2})(x - \sqrt{-2}).$$

He began by assuming $\gcd(x + \sqrt{-2}, x - \sqrt{-2}) = 1$, and we showed in Section 2.9 that this assumption is valid when $x$ is odd.

A similar problem arises with the equation

$$y^3 = x^2 + 5 = (x + \sqrt{-5})(x - \sqrt{-5}),$$

where we would like to know that $\gcd\left(x + \sqrt{-5}, x - \sqrt{-5}\right) = 1$. This is false when $x$ is odd – take $x = 5$, for example – but we can show it is true when $x$ is even. Since $\mathbb{Z}\left[\sqrt{-5}\right]$ is not a principal ideal domain, this means showing that no prime *ideal* divides both $x + \sqrt{-5}$ and $x - \sqrt{-5}$, in the sense of dividing the ideals $(x + \sqrt{-5})$ and $\left(x - \sqrt{-5}\right)$.

1. Suppose $P$ is a prime ideal that divides $x + \sqrt{-5}$ and $x - \sqrt{-5}$. Explain why $2x \in P$ and $2\sqrt{-5} \in P$.
2. Deduce that either $2 \in P$ or else $x \in P$ and $\sqrt{-5} \in P$.
3. If $2 \in P$, deduce that 2 divides $x^2 + 5 = y^3$, so $y$ is even and hence $x$ is odd. Show that this leads to a contradiction.
4. If $x, \sqrt{-5} \in P$ then $5 \in P$ (can you see why?) and hence 5 divides $x$ (otherwise $P = (1)$, which implies it is not prime). Show that this also leads to a contradiction.

Now, to cut a long story short, we can reach a conclusion like Euler did: if $y^3 = x^2 + 5$, then $x + \sqrt{-5}$ is a cube $\left(a + b\sqrt{-5}\right)^3$ for some $a, b \in \mathbb{Z}$.

5. Show that $x + \sqrt{-5} = (a + b\sqrt{-5})^3$ for $a, b \in \mathbb{Z}$ leads to a contradiction, and hence that $y^3 = x^2 + 5$ has no solution in ordinary integers.

## 9.8 Discussion

As I said once before, in the introduction to (Dedekind, 1996, p. 3):

> Dedekind's invention of ideals in the 1870s was a major turning point in the development of algebra. His aim was to apply ideals to number theory, but to do this he had to build the whole framework of commutative algebra: fields, rings, modules and vector spaces. These concepts, together with groups, were to form the core of the future abstract algebra. At the same time he created algebraic number theory, which became the temporary home of algebra while its core concepts were growing up.

Dedekind's theory of ideals was reworked by Dedekind himself – for example, fractional ideals first appear in Dedekind (1894) – then by Noether (1921) and Noether (1926), where the Noetherian and integrality components of the Dedekind domain concept are identified. Noether's ideas were polished in van der Waerden (1931), from which they spread widely. After van der Waerden, it becomes difficult to trace their continued evolution, which includes greater emphasis on the maximality statement[3] of the Noetherian property, as

---

[3] In the paper that introduces the term "Noetherian," Chevalley (1943) uses the same definition of this property as Emmy Noether – that each ideal is finitely generated – but he *calls* it a

we have seen in this chapter. The book of Zariski and Samuel (1958) was influential, and its offshoot Samuel (1970) was particularly helpful to me in preparing the present book.

### 9.8.1 Algebraic Function Theory

In the Preface I mentioned the "second life" of Dedekind domains in algebraic geometry. This stems from the paper Dedekind and Weber (1882), which may be seen in English translation as Dedekind and Weber (2012). It is impossible to do justice to algebraic geometry in a book of this size, but the story begins with an analogy between algebraic number fields and algebraic function fields that was sketched in the exercises to Section 8.5. The following table summarizes the analogy.

| Number Fields | Function Fields |
|---|---|
| domain $\mathbb{Z}$ | domain $\mathbb{C}[x]$ |
| field $\mathbb{Q}$ of fractions | field $\mathbb{C}(x)$ of fractions |
| algebraic number $x$ satisfies | algebraic function $y$ satisfies |
| $a_n x^n + \cdots + a_0 = 0$ with | $a_n y^n + \cdots + a_0 = 0$ with |
| $a_0, \ldots, a_n \in \mathbb{Z}$ | $a_0, \ldots, a_n \in \mathbb{C}[x]$ |
| algebraic number field $E$ is | algebraic function field $F$ is |
| extension $E \supseteq \mathbb{Q}$ of finite dimension | extension $F \supseteq \mathbb{C}(x)$ of finite dimension |
| integers of $E$ have $a_n = 1$ | integral functions of $F$ have $a_n = 1$ |
| and they form a Dedekind domain | and they form a Dedekind domain |

Before the 19th century, algebraic functions came up mainly in calculus, where the problem of *integrating* them was observed to be difficult – and to be connected with problems of algebra and number theory. Even the class $\mathbb{C}(x)$ of rational functions posed a problem. A rational function $f(x) = p(x)/q(x)$ is a quotient of polynomials, which may be integrated by the method of partial fractions provided $q(x)$ has a factorization into linear factors. *But the existence of such a factorization depends on the fundamental theorem of algebra.* In fact, the problem of integrating rational functions was the main motive for proving the fundamental theorem of algebra in the first place.

When the rational functions are extended by even the square root function (an algebraic function $y$ satisfying $y^2 - x = 0$), the difficulties multiply. A few

---

"maximal condition." Evidently, by this time, the equivalence of finite generation with existence of a maximal element in any set of ideals went without saying.

such functions can be integrated by *rational change of variable*. For example, we can find

$$\int \frac{dx}{\sqrt{1-x^2}} \quad \text{by making the substitution} \quad x = \frac{1-t^2}{1+t^2},$$

which "rationalizes" the integral to

$$\int \frac{-2dt}{1+t^2}.$$

The reason behind this seemingly miraculous simplification is that the function $y = \sqrt{1-x^2}$ satisfies the equation $x^2 + y^2 = 1$ for the unit circle. We are able to rationalize the integral because we are able to "rationalize the circle" by the parametric equations

$$x = \frac{1-t^2}{1+t^2}, \quad y = \frac{2t}{1+t^2},$$

which are those found in our search for Pythagorean triples in Section 2.1. There we were interested in rational values of $t$, but it is easy to see that these functions $x$ and $y$ of $t$ satisfy $x^2 + y^2 = 1$ for all complex values of $t \neq \pm i$.

However, rational substitution fails to work for the integral

$$\int \frac{dx}{\sqrt{1-x^4}}$$

because it fails to work for the curve

$$y = \sqrt{1-x^4}, \quad \text{or} \quad y^2 = 1 - x^4.$$

As we saw in the exercises to Section 2.2, the reason for this is also related to number theory, namely, a theorem of Fermat (1670) implying that the equation $y^2 = 1-x^4$ has no rational *number* solutions. The proof goes from the formula for Pythagorean triples (Section 2.1) to the result that $c^2 = a^4 - b^4$ has no solution in integers. It then follows by division that $y^2 = 1-x^4$ has no solution in rational numbers. A completely analogous proof works with polynomials in place of integers, since polynomials enjoy unique prime factorization, so we can conclude that $y^2 = 1 - x^4$ has no solution in rational functions.

These two examples suggest that difficulties in integration can be traced to the nature of curves: particularly the fact that some curves can be parameterized by rational functions and others cannot. The first to really grasp this idea was Riemann, around 1850. However, long before this realization dawned, Abel (1826b) had proved an amazing theorem measuring the complexity of integrals of algebraic functions by a number he called the *genus*. Abel's theorem is too complicated to explain here, but we can explain one of Riemann's discoveries, which is that genus has a *topological* interpretation.

Figure 9.1 Bernhard Riemann (1826–1866) (public domain).

### 9.8.2 Algebraic Curves

The equations that define **algebraic functions** (of one variable),

$$a_n(x)y^n + \cdots + a_1(x)y + a_0(x) = 0, \quad \text{where} \quad a_0(x)\ldots,a_n(x) \in \mathbb{C}[x], \quad (*)$$

are simply polynomial equations in the two variables $x, y$, and hence they also define **algebraic curves**. To ensure that every polynomial equation has a root, it is usual to let $x$ and $y$ take all complex values, so the values of $x$ form the plane $\mathbb{C}$. (It is also usual to include the value $\infty$, which closes the plane to a sphere, but we will focus on the finite values of $x$ and $y$ for now.) The values of $y$ form another copy of this plane.

Riemann's great idea was to view the $y$-plane as a **covering** of the $x$-plane, with the finitely many values of $y$ that satisfy (*) for a given value $x_0$ lying "above" $x_0$. Apart from finitely many exceptional values of $x_0$, called **branch points**, the covering in a neighborhood of $x_0$ looks like a stack of parallel planes, called **sheets** of the covering. Above a branch point, two or more sheets spiral around a common point. The simplest case, shown in Figure 9.2, is where the two sheets of the covering for $y = \sqrt{x}$ meet over $x = 0$. Over every nonzero point $x$ there are two points of the covering, corresponding to $+\sqrt{x}$ and $-\sqrt{x}$. This picture, from Neumann (1865), is misleading in one respect. Trying to view the relation between the $x$-plane and $y$-plane in three dimensions forces us to create a line of intersection of the two sheets. In reality, the sheets of the $y$-plane meet *only* at the single point $y = 0$, and the

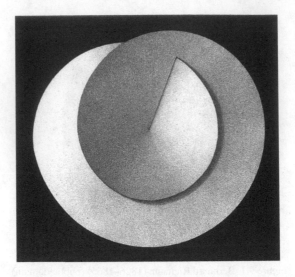

Figure 9.2  Picture of a branch point. From *Vorlesungen über Riemann's Theorie der Abel'schen Integrale* by Carl Neumann, B. G. Teubner, 1865.

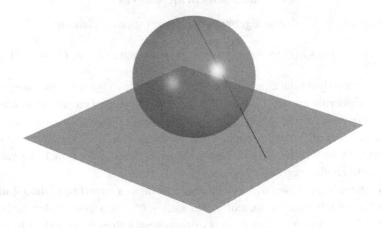

Figure 9.3  Correspondence between the sphere and $\mathbb{C} \cup \{\infty\}$.

neighborhood of this point is a disk, like the neighborhood of any other point on the $y$-plane.

When we include $\infty$ among the values of $x$, it is natural to view the collection of $x$-values as a *sphere*, often called the **Riemann sphere**. This is explained by Figure 9.3, which shows how a sphere is mapped on to a plane by projection from its "north pole."

Figure 9.4 Riemann surfaces of genus 1, 2, 3.

Under this map, points near to the north pole are sent to distant points of the plane, so the point $\infty$ in the plane naturally corresponds to the north pole itself. Thus, if we allow $x$ and $y$ to take the value $\infty$, the relationship (*) between $x$ and $y$ becomes a covering of a sphere by a sphere with finitely many branch points, called a **Riemann surface**. For example, the covering for $y = \sqrt{x}$ now has *two* branch points—the other one at the point $x = \infty$.

Because sheets fuse at branch points, the surface of $y$ values is not necessarily a sphere. In fact, it can also take the forms shown in Figure 9.4. By a form, we now mean *topological* form, where any two surfaces in continuous one-to-one correspondence are considered to be of the "same form." It is intuitively plausible, and can be proved rigorously, that the number of "holes" in the surface characterizes its topological form. The number of "holes" is called the **genus** because, incredibly, it is the *same number* as Abel's genus, when his theorem is suitably interpreted as a fact about algebraic functions.

One of the consequences of this discovery of Riemann is that the curves with genus 0 (that is, of spherical form) are precisely those that can be parameterized by rational functions.

### 9.8.3 Construction of a Riemann Surface from a Function Field

The sudden appearance of Riemann surfaces, and their topology, in the theory of algebraic functions, caught 19th-century mathematicians by surprise. They had no rigorous framework for handling these concepts, and Riemann himself could back up his amazing insights only by intuition, which even included appeals to physics. Also, Riemann died before he was 40, so he did not have time to reflect, revise, and put his discoveries on a rigorous foundation.

It fell to Dedekind and Weber to edit Riemann's collected works in the 1870s. They determined to build a rigorous foundation for Riemann's theory, by turning Dedekind's approach to algebraic numbers towards algebraic functions. As we have seen at the beginning of this section, there is a strong analogy between algebraic numbers and algebraic functions. But there are also special concepts in algebraic function theory with no apparent counterpart in algebraic number theory, such as Riemann surfaces.

One of the triumphs of the Dedekind–Weber theory was to build a Riemann surface from an algebraic function field. The idea is this: *a point gives values to functions*. Therefore, an assignment of values to the functions in a field can be regarded as a **point**, provided values are assigned consistently, in the following sense:

- The value given to the constant function $c$ is $c$.
- If $f$ and $g$ get values $a$ and $b$, then $f + g$ gets value $a + b$.
- If $f$ and $g$ get values $a$ and $b$, then $f - g$ gets value $a - b$.
- If $f$ and $g$ get values $a$ and $b$, then $f \cdot g$ gets value $a \cdot b$.
- If $f$ and $g$ get values $a$ and $b$, then $f/g$ gets value $a/b$.

This is a brilliant start, but so far we have only a cloud of points. How can this cloud be regarded as a "surface"? In particular, how can it have a genus?

This problem is solved with the help of branch points. In Riemann's theory, the genus of a surface is completely determined by the nature of its branch points (by a formula known as the **Riemann–Hurwitz formula**). The Dedekind–Weber theory finds the nature of branch points algebraically. For example, we can detect the branch point at $x = 0$ for $y = \sqrt{x}$, and the fact that it involves two sheets, from the equation $y^2 - x = 0$. This makes it possible to *define* genus algebraically. With this definition, Dedekind and Weber were able to give a purely algebraic proof of Abel's theorem.

There is much more, but perhaps this is enough to give a glimpse of the Dedekind–Weber theory of algebraic curves, a theory summed up in the *History of Algebraic Geometry* by (Dieudonné, 1985, p. 29), as follows.

> Dedekind and Weber propose to give algebraic proofs of all of Riemann's algebraic theorems. But their remarkable originality (which in all the history of algebraic geometry is only scarcely surpassed by that of Riemann) leads them to introduce a series of ideas that will become fundamental in the modern era.

# References

Abel, N. H. (1826a). Démonstration de l'impossibilité de la résolution algébrique des équations générales qui passent le quatrième degré. *Journal für die reine und angewandte Mathematik 1*, 65–84. *Oeuvres Complètes* 1: 66–87.

Abel, N. H. (1826b). Mémoire sur une propiété générale d'une classe très étendue de fonctions transcendantes. *Mémoire des Savants Étrangers, 1841 7*, 176–264. *Oeuvres Complètes* 2: 145–211.

Artin, E. (1942). *Galois Theory*, Notre Dame Mathematical Lectures, no. 2. University of Notre Dame, Notre Dame, IN. Edited and supplemented with a section on applications by Arthur N. Milgram.

Bashmakova, I. G. and G. S. Smirnova (2000). *The Beginnings and Evolution of Algebra*, Volume 23 of *The Dolciani Mathematical Expositions*. Mathematical Association of America, Washington, DC. Translated from the Russian by Abe Shenitzer with the editorial assistance of David A. Cox.

Bernoulli, J. (1704). Positionum de seriebus infinitis ... pars quinta. In his *Werke*, 4, 127–147.

Blass, A. (1984). Existence of bases implies the axiom of choice. In *Axiomatic Set Theory (Boulder, Colorado, 1983)*, Volume 31 of Contemporary Mathematics. American Mathematical Society, Providence, RI, 31–33.

Cantor, G. (1874). Über eine Eigenschaft des Inbegriffes aller reellen algebraischen Zahlen. *Journal für reine und angewandte Mathematik 77*, 258–262. English translation by W. Ewald in Ewald (1996), Vol. II, 840–843.

Cantor, G. (1883). *Grundlagen einer allgemeinen Mannigfaltigkeitslehre*. Teubner, Leipzig. In his *Gesammelte Abhandlungen*, Springer, Berlin, 165–204. English translation by W. Ewald in Ewald (1996), Vol. II, 878–919.

Cantor, G. (1891). Über eine elementare Frage der Mannigfaltigkeitslehre. *Jahresbericht deutschen Mathematiker-Vereinigung 1*, 75–78. English translation by W. Ewald in Ewald (1996), Vol. II, 920–922.

Cardano, G. (1545). *Ars Magna or The Rules of Algebra*. Dover Publications, Inc., New York, 1993. Translated from the Latin and edited by T. Richard Witmer, With a foreword by Oystein Ore.

Cauchy, A.-L. (1847). Mémoire sur la théorie des équivalences algébriques, substituée à la théorie des imaginaires. In *Exercises d'analyse et de physique mathématique*, Tome 4, Bachelier, Paris, 87–110.

Cayley, A. (1855). Remarques sur la notation des fonctions algébriques. *Journal für die reine und angewandte Mathematik 50*, 282–285.

Chevalley, C. (1943). On the theory of local rings. *Annals of Mathematics. Second Series 44*, 690–708.

Cohen, P. (1963). The independence of the continuum hypothesis. *Proceedings of the National Academy of Sciences of the United States of America 50*, 1143–1148.

Cramer, G. (1750). *Introduction à l'analyse des lignes courbes algébriques*. Geneva.

Dedekind, R. (1871). Supplement X. In Dirichlet's *Vorlesungen über Zahlentheorie*, 2nd ed., Vieweg 1871.

Dedekind, R. (1894). Supplement XI. In Dirichlet's *Vorlesungen über Zahlentheorie*, 4th ed., Vieweg 1894.

Dedekind, R. (1996). *Theory of Algebraic Integers*. Cambridge Mathematical Library. Cambridge University Press, Cambridge. Translated from the 1877 French original and with an introduction by John Stillwell.

Dedekind, R. and H. Weber (1882). Theorie der algebraischen Functionen einer Veränderlichen. *Journal für die reine und angewandte Mathematik 92*, 181–291.

Dedekind, R. and H. Weber (2012). *Theory of Algebraic Functions of One Variable*, Volume 39 of *History of Mathematics*. American Mathematical Society, Providence, RI; London Mathematical Society, London. Translated from the 1882 German original and with an introduction, bibliography and index by John Stillwell.

Descartes, R. (1637). *The Geometry of René Descartes. (With a Facsimile of the First edition, 1637.)*. Dover Publications Inc., New York. Translated by David Eugene Smith and Marcia L. Latham, 1954.

Dieudonné, J. (1985). *History of Algebraic Geometry: An Outline of the History and Development of Algebraic Geometry*. Wadsworth Mathematics Series. Wadsworth International Group, Belmont, CA. Translated from the French by Judith D. Sally.

Dirichlet, P. G. L. (1863). *Vorlesungen über Zahlentheorie*. F. Vieweg und Sohn, Braunschweig. English translation *Lectures on Number Theory*, with Supplements by R. Dedekind, translated from the German and with an introduction by John Stillwell, American Mathematical Society, Providence, RI, 1999.

Eisenstein, G. (1850). Über einige allgemeine Eigenschaften der Gleichung, von welcher die Theorie der ganzen Lemniscate abhängt. *Journal für die reine und angewandte Mathematik 39*, 556–619.

Euler, L. (1747). Theoremata circa divisors numerorum. *Novi commentarii academiae scientiarum Petropolitanae 1*, 20–48.

Euler, L. (1748). Letter to Goldbach, 4 May 1748. In Fuss (1968), **1**, 450–455.

Euler, L. (1770). *Elements of Algebra*. Springer-Verlag, New York, 1984. Translated from the German by John Hewlett. Reprint of the 1840 edition, with an introduction by C. Truesdell.

Ewald, W. (1996). *From Kant to Hilbert: A Source Book in the Foundations of Mathematics. Vols. I, II*. The Clarendon Press, Oxford University Press, New York.

Fermat, P. (1640). Letter to Mersenne, 25 December 1640. *Œuvres 2*: 212.

Fermat, P. (1654). Letter to Pascal, 25 September 1654. *Œuvres* 2: 310–314.

Fermat, P. (1657). Letter to Frenicle, February 1657. *Œuvres* 2: 333–334.

Fermat, P. (1670). Observations sur Diophante. *Œuvres* 3: 241–276.

Fibonacci, L. P. (1225). *Liber quadratorum*. Academic Press, Inc., Boston, MA, 1987. Translated from the Latin as *The Book of Squares* and with a preface, introduction and commentaries by L. E. Sigler.

Friedman, H. M., S. G. Simpson, and R. L. Smith (1983). Countable algebra and set existence axioms. *Annals of Pure and Applied Logic* 25(2), 141–181.

Fuss, P.-H. (1968). *Correspondance mathématique et physique de quelques célèbres géomètres du XVIIIème siècle. Tomes I, II.* Johnson Reprint Corp., New York. Reprint of the Euler correspondence originally published by l'Académie Impériale des Sciences de Saint-Pétersbourg. *The Sources of Science*, No. 35.

Gauss, C. F. (1801). *Disquisitiones Arithmeticae*. Springer-Verlag, New York, 1986. Translated and with a preface by Arthur A. Clarke. Revised by William C. Waterhouse, Cornelius Greither and A. W. Grootendorst and with a preface by Waterhouse.

Gauss, C. F. (1816). Demonstratio nova altera theorematis omnem functionem algebraicum rationalem integram unius variabilis in factores reales primi vel secundi gradus resolvi posse. *Commentationes Societatis Regiae Scientiarum Gottingensis recentiores 3*, 107–142. In his *Werke* 3: 31–56.

Gödel, K. (1938). The consistency of the axiom of choice and of the generalized continuum-hypothesis. *Proceedings of the National Academy of Sciences of the United States of America 24*, 556–557.

Grassmann, H. (1844). *Die lineale Ausdehnungslehre*. Otto Wigand, Leipzig. English translation in Grassmann (1995), 1–312.

Grassmann, H. (1861). *Lehrbuch der Arithmetic*. Enslin, Berlin.

Grassmann, H. (1862). *Die Ausdehnungslehre*. Enslin, Berlin. English translation of 1896 edition in Grassmann (2000).

Grassmann, H. (1995). *A New Branch of Mathematics*. Open Court Publishing Co., Chicago, IL. The *Ausdehnungslehre* of 1844 and other works, Translated from the German and with a note by Lloyd C. Kannenberg. With a foreword by Albert C. Lewis.

Grassmann, H. (2000). *Extension Theory*. American Mathematical Society, Providence, RI; London Mathematical Society, London. Translated from the 1896 German original and with a foreword, editorial notes and supplementary notes by Lloyd C. Kannenberg.

Gray, J. (2018). *A History of Abstract Algebra*. Springer, Cham, Switzerland.

Halmos, P. R. (1942). *Finite Dimensional Vector Spaces*. Annals of Mathematics Studies, no. 7. Princeton University Press, Princeton, NJ.

Hamel, G. (1905). Eine Basis aller Zahlen und die unstetigen Lösungen der Funktionalgleichung $f(x + y) = f(x) + f(y)$. *Mathematische Annalen 60*, 459–462.

Heath, T. L. (1964). *Diophantus of Alexandria: A Study in the History of Greek Algebra*. Second edition. With a supplement containing an account of Fermat's theorems and problems connected with Diophantine analysis and some solutions of Diophantine problems by Euler. Dover Publications, Inc., New York.

Hilbert, D. (1890). Über die Theorie der algebraischen formen. *Mathematische Annalen 36*, 473–534.

Hilbert, D. (1897). *Zahlbericht*. Springer-Verlag, Berlin, 1998. Translated from the German as *The Theory of Algebraic Number Fields* and with a preface by Iain T. Adamson, With an introduction by Franz Lemmermeyer and Norbert Schappacher.

Hodges, W. (1979). Krull implies Zorn. *The Journal of the London Mathematical Society. Second Series 19*(2), 285–287.

Jacobson, N. (1985). *Basic Algebra I* (Second ed.). W. H. Freeman and Company, New York.

Kaplansky, I. (1991). Reminiscences. In *Paul Halmos. Celebrating 50 Years of Mathematics*, 297–304. New York: Springer-Verlag.

Katz, V. J. and K. H. Parshall (2014). *Taming the Unknown: A History of Algebra from Antiquity to the Early Twentieth Century*. Princeton University Press, Princeton, NJ.

Knobloch, E. (2013). Leibniz's theory of elimination and determinants. In *Seki, Founder of Modern Mathematics in Japan*, Volume 39 of *Springer Proceedings in Mathematics & Statistics*, 229–244. Springer, Tokyo.

Kolmogorov, A. N. and A. P. Yushkevich (Eds.) (2001). *Mathematics of the 19th Century* (revised ed.). Birkhäuser Verlag, Basel. Mathematical logic. Algebra. Number theory. Probability theory, Translated from the 1978 Russian original by A. Shenitzer, H. Grant and O. B. Sheinin, Translation edited by Shenitzer.

Kronecker, L. (1870). Auseinandersetzung einige Eigenschaften der Klassenzahl idealer complexer Zahlen. *Monatsbericht der Königlich Preussischen Akademie der Wissenschaften zu Berlin*, 881–889. In his *Werke* 1, 271–282.

Kronecker, L. (1887). Ein Fundamentalsatz der allgemeinen Arithmetik. *Journal für die reine und angewandte Mathematik 100*, 490–510.

Krull, W. (1928). Zur Theorie der allgemeinen Zahlringe. *Mathematischen Annalen 99*, 51–70.

Krull, W. (1929). Idealtheorie in Ringen ohne Endlichkeitsbedingungen. *Mathematischen Annalen 101*, 729–744.

Krull, W. (1935). *Idealtheorie*. Springer, Berlin.

Kummer, E. E. (1844). De numeris complexis, qui radicibus unitatis et numeris realibus constant. *Gratulationschrift der Universität Breslau zur Jubelfeier der Universität Königsberg*. Also in Kummer (1975), vol. 1, 165–192.

Kummer, E. E. (1975). *Collected Papers*. Volume 1, *Contributions to Number Theory*. Springer-Verlag, Berlin-New York. Edited and with an introduction by André Weil.

Kuratowski, C. (1922). Une méthode d'elimination des nombres transfinis des raisonnements mathématiques. *Fundamenta Mathematicae 3*, 76–108.

Lagrange, J. L. (1768). Solution d'un problème d'arithmétique. *Miscellanea Taurinensia 4*, 19ff. In his *Œuvres* 1: 671–731.

Lagrange, J. L. (1773). Recherches d'arithmétique. *Nouveaux mémoires de l'Académie royale des sciences et belles-lettres de Berlin*, 265ff. In his *Œuvres* 3: 695–795.

Laplace, P.-S. (1772). Researches sur le calcul intégral et sur le systéme du monde. *Histoire de l'Académie Royale des Sciences, Paris*, 267–376.

Muir, T. (1960). *The Theory of Determinants in the Historical Order of Development.* Dover Publications, Inc., New York. Four volumes bound as two.

Neumann, C. (1865). *Vorlesungen über Riemann's Theorie der Abelschen Integralen.* Leipzig: Teubner.

Noether, E. (1921). Idealtheorie in Ringbereichen. *Mathematischen Annalen 83,* 24–66.

Noether, E. (1926). Abstrakter Aufbau der Idealtheorie in algebraischen Zahl- und Funktionenkörpern. *Mathematischen Annalen 96,* 26–61.

Pascal, B. (1654). Traité du triangle arithmétique, avec quelques autres petits traités sur la même manière. English translation in *Great Books of the Western World,* Encyclopedia Britannica, London, 1952, 447–473.

Peano, G. (1888). *Calcolo Geometrico secondo l'Ausdehnungslehre di H. Grassmann, preceduto dalle operazioni della logica deduttiva.* Bocca, Turin. English translation in Peano (2000).

Peano, G. (2000). *Geometric Calculus: According to the* Ausdehnungslehre *of H. Grassmann.* Birkhäuser Boston, Inc., Boston, MA. Translated from the Italian by Lloyd C. Kannenberg.

Rothe, H. A. (1800). Ueber Permutationen, in Beziehung auf die Stellen ihrer Elemente. Anwendung der daraus abgeleiteten Sätze auf das Eliminationsproblem. In Hindenburg, Carl, ed., *Sammlung Combinatorisch-Analytischer Abhandlungen,* 263–305, Bey G. Fleischer dem jüngern, 1800.

Rotman, J. J. (2015). *Advanced Modern Algebra, Part 1* (Third ed.), Volume 165 of *Graduate Studies in Mathematics.* American Mathematical Society, Providence, RI.

Samuel, P. (1970). *Algebraic Theory of Numbers.* Houghton Mifflin Co., Boston, MA. Translated from the French by Allan J. Silberger.

Simpson, S. G. (2009). *Subsystems of Second Order Arithmetic* (Second ed.). Perspectives in Logic. Cambridge University Press, Cambridge; Association for Symbolic Logic, Poughkeepsie, NY.

Smith, D. E. (1959). *A Source Book in Mathematics.* 2 vols. Dover Publications Inc., New York

Solovay, R. M. (1970). A model of set-theory in which every set of reals is Lebesgue measurable. *Annals of Mathematics. Second Series 92,* 1–56.

Steinitz, E. (1910). Algebraische Theorie der Körper. *Journal für die reine und angewandte Mathematik 137,* 167–302.

Stillwell, J. (1994). *Elements of Algebra: Geometry, Numbers, Equations.* Undergraduate Texts in Mathematics. Springer-Verlag, New York.

Stillwell, J. (2018). *Reverse Mathematics.* Princeton University Press, Princeton, NJ.

van der Waerden, B. L. (1931). *Moderne Algebra II* Teil. Unter Benutzung von Vorlesungen von E. Artin und E. Noether. Die Grundlehren der mathematischen Wissenschaften in Einzeldarstellungen mit besonderer Berücksichtigung der Anwendungsgebiete. Bd. 34. Julius Springer, Berlin, vii, 216 S. (1931).

van Heijenoort, J. (1967). *From Frege to Gödel: A Source Book in Mathematical Logic, 1879–1931.* Harvard University Press, Cambridge, MA.

Vandermonde, A.-T. (1771). Mémoire sur la résolution des équations. *Mémoires de l'Academie royale des sciences à Paris,* 365–416.

Viète, F. (1591). De aequationum recognitione et emendatione. In his *Opera*, 82–162. English translation in Viète (1983).

Viète, F. (1615). Ad angularium sectionum analyticen theoremata. In his *Opera*, 287–304.

Viète, F. (1983). *The Analytic Art: Nine studies in algebra, geometry and trigonometry from the* Opus restitutae mathematicae analyseos, seu algebrâ novâ. Kent State University Press, Kent, OH. Translated from the Latin and with an introduction by T. Richard Witmer.

Vitali, G. (1905). *Sul problema della misura dei gruppi di punti di una retta*. Bologna.

Weil, A. (1974). Two lectures on number theory, past and present. *L'Enseignement Mathématique. Revue Internationale. IIe Série 20*, 87–110.

Weil, A. (1975). Introduction to Kummer (1975).

Weil, A. (1984). *Number Theory: An Approach through History, from Hammurapi to Legendre*. Birkhäuser Boston, Inc., Boston, MA.

Wessel, C. (1797). Om Directionens analytiske Betegning, et Forsøg anvendt fornemmelig til plane og sphæriske Polygoners Opløsning. *Danske Selsk. Skr. N. Samml. 5*. English translation in Smith (1959), vol. 1, 55–66.

Wiles, A. (1995). Modular elliptic curves and Fermat's last theorem. *Annals of Mathematics (2) 141*(3), 443–551.

Ycart, B. (2013). A case of mathematical eponymy: The Vandermonde determinant. *Revue d'Histoire des Mathématiques 19*(1), 43–77.

Zariski, O. and P. Samuel (1958). *Commutative Algebra, Volume I*. The University Series in Higher Mathematics. D. Van Nostrand Company, Inc., Princeton, NJ. With the cooperation of I. S. Cohen.

Zermelo, E. (1904). Beweis dass jede Menge wohlgeordnet werden kann. *Mathematische Annalen 59*, 514–516. English translation in van Heijenoort (1967).

Zorn, M. (1935). A remark on method in transfinite algebra. *Bulletin of the American Mathematical Society 41*, 667–670.

# Index

Abel, Niels Henrik, 100
  theorem on genus, 206, 210
Abelian group
  additive notation, 73
  as a $\mathbb{Z}$-module, 172
  axioms, 69
  definition, 69
  finite, 68
  of fractional ideals, 199
absolute value
  of complex number, 39
  multiplicative property, 40
AC. *See* axiom of choice, 122
$ACA_0$, 125
ACC. *See* ascending chain condition, 114
additive notation, 73
al-Khwarizmi, 26
algebra
  commutative, xii
  linear, xii
  origin of word, 26
algebraic
  curve, 169, 207
  field extension, 90
  function, 184
    definition, 207
    integral, 184
    ring, 188
  function field, 184, 188
    gives Riemann surface, 210
  function theory, 205
  geometry, 27, 57, 188
  integer, 11, 14, 31, 53
    defined by Dedekind, 31, 67

    definition, 93
    factorization of, 95
    is integer over $\mathbb{Z}$, 183
    second definition, 96
    sum and product, 132
    used by Euler, 33
  number, 86, 101
    as fraction of integers, 175
    as matrix, 136
    norm via det, 138, 161
    sum and product, 132
    trace, 161
  number field, 30, 78, 86
    conjugates in, 88
    embedded in $\mathbb{C}$, 142
    integer of, 32, 93, 94, 174
    Kronecker approach, 31
arithmetic comprehension axiom, 125
  algebraic equivalents, 125
Artin, Emil, 150
ascending chain, 110
ascending chain condition, 114
  as "divisor chain theorem,", 191
  and finite generation, 117
  in Hilbert's basis theorem, 117
  in a PID, 111
axiom of choice, 115, 122
  for algebraists, 123
  and existence of basis, 128
  and maximal ideals, 123
  and nonmeasurability, 123
  and vector space basis, 123
  and well-ordering, 120, 123
  and Zorn's lemma, 123

217

axioms
  Abelian group, 69
  for determinant, 150
  field, 19, 79
  of infinity, 123
  Peano (for arithmetic), 25, 124
  ring, 15, 79
  for set existence, 124
  vector space, 127, 146

basis
  and axiom of choice, 128
  of field extension, 90
  finite, 128
  Hamel, 129
  independence, 136
    of charpoly, 149, 159
    of det, 136, 149, 159
    of norm, 149
    of trace, 136, 149
  integral, 149, 168, 175
    definition, 176
    existence, 176
  of module, 173, 180
  of vector space, 126
  for $\mathbb{Q}(\alpha)$, 91
Bernoulli, Jakob, 38
Bhaskara II, 25
Binet, Jacques, 170
binomial
  coefficient, 28
    in Chinese mathematics, 28
    and Fermat's little theorem, 29
  theorem, 28
Bourbaki, 123
Brahmagupta, 25
  identity, 58, 66
branch point, 207

$\mathbb{C}$, 70, 99
  Cauchy construction, 102
calculus, 27
  of algebraic functions, 205
  and FTA, 205
Cantor, Georg, 101
  assumed well-ordering, 123
  discovered uncountability, 121
  pairing function, 120
Cardano, Gerolamo, 26
Cauchy, Augustin-Louis
  construction of $\mathbb{C}$, 102
  rebuilt determinant theory, 170

Cayley, Arthur, 148
characteristic polynomial, xii, 149
  definition, 159
  is basis-independent, 149, 159
  is minimal, 162
charpoly. *See* characteristic polynomial, 149
Chinese remainder theorem, 71
choice function, 122
class group, 67, 174
class number, 64
cofactor, 153
Cohen, Paul, 123
commutative algebra, xii
complex number, 39, 55
  absolute value, 39
  multiplication, 40
composition of forms, 66
  associativity, 67
  by Dedekind, 76
congruence, 19
  class, 19
    mod a module, 173
    mod a polynomial, 85, 88
    mod an ideal, 109
  mod a module, 173
  mod a polynomial, 84
  mod an ideal, 104, 109
conjugate, 13
  in algebraic number field, 88, 143
  in cyclotomic field, 164
  of Gaussian integer, 41
  ideal, 191
  multiplicative property, 13, 41
  roots of minimal polynomial, 143
  in $\mathbb{Z}[\sqrt{-2}]$, 51
content, 96
continued fraction, 4
  and Euclidean algorithm, 4
  infinite, 12
countability, 119
Cramer, Gabriel, 145
  rule, 158, 169
cyclotomic
  field, 164, 178
  integer, 102, 179
  polynomial, 98
  reason for name, 98

Dedekind, Richard, xii
  algebraic function theory, 188
  composition of forms, 76
  defined algebraic integers, 31, 103

domain, xii
ideals, 75
induction, 25
introduced fractional ideals, 204
module concept, 173
norm concept, 58
on norm and discriminant, 168
on Vandermonde deteminant, 169
praise for Kummer, 75
product theorem, 131
realized ideal numbers, 104
supplements to Dirichlet, 76
theory of algebraic integers, 67, 76, 186
theory of ideals, 68
Dedekind domain, 189, 197
in algebraic geometry, 205
definition, 193
Dedekind ring. *See* Dedekind domain, 193
degree, 84
as dimension, 133
function, 17
of zero polynomial, 16, 84
del Ferro, Scipione, 26
dependent choice, 116, 117, 123
derivative
criterion for multiple root, 140
of polynomial, 139
product rule, 140
Descartes, René, 27
algebraic geometry, 57
factor theorem, 22
det. *See* determinant, 58
determinant, xii, 126
axioms, 150
cofactor, 153
column expansion, 155
computing rules, 154
criterion for linear dependence, 156
criterion for nonzero solution, 94, 157
definition of charpoly, 149
definition of norm, 138, 149
existence, 153
is basis-independent, 136, 149, 159
is coefficient of charpoly, 160
and linear equations, 145, 156
multiplicative property, 58, 66, 149
proof, 152
and permutations, 152
row expansion, 155
in study of algebraic integers, 94, 132
theory, 149
history, 169

of transpose, 155
uniqueness, 152
Vandermonde, 166
diagonal argument, 122
differential geometry, 27
dimension
as degree, 133
of extension field, 90
notation, 131
finite, 128
of free module, 180
invariance under isomorphism, 134
relative, 142
of a subspace, 132
of vector space, 126, 128
Diophantine equation, 33
linear, 6
of higher degree, 7
Diophantus, 26, 33
*Arithmetica*, 38, 53, 55
chord method, 34, 54
identity, 39, 55
and multiplicative property, 39
and sums of squares, 38, 59
as composition of forms, 66
tangent method, 54, 55
direct product
of groups, 69
of rings, 71
direct sum, 74
Dirichlet, Peter Lejeune, 31
simplified Gauss, 76
discriminant, xii, 164
criterion for a basis, 167
definition, 165
Lagrange, 167
of quadratic form, 59, 64
test for basis, 149
of transformed basis, 177
trace, in terms of, 165
of transformed basis, 166
used to find module bases, 171, 176
division
in field, 19
ideal, 190
and product of ideals, 200
property, 2
with remainder, 2
in a ring, 15
division property, 2
fails in $\mathbb{Z}[\sqrt{-3}]$, 52

division property (Cont.)
  fails in $\mathbb{Z}[\sqrt{-5}]$, 63
  for polynomials, 16
  in $\mathbb{Z}[i]$, 42, 44
  in $\mathbb{Z}[\sqrt{-2}]$, 45, 52
  in $\mathbb{Z}[\zeta]$, 53
division with remainder
  and continued fraction, 4
  and matrices, 5
  in $\mathbb{N}$, 2
divisor, 2
domain, 82
  Dedekind, 189
    definition, 193
  definition, 82
  Euclidean, 17
  finite, 196
  fractions of, 82
  integrally closed, 184
  Noetherian, 194
  principal ideal, 8, 10, 108, 173, 190
duplication of the cube, 133

Eisenstein, Gotthold, 31
  and algebraic integers, 102
  irreducibility criterion, 97
equation
  cubic, 26
  Diophantine, 33
    of higher degree, 7
    linear, 6, 25
    quadratic, 25
  Pell, 13
  polynomial, xi
equivalence
  class, 64, 79
    disjointness, 81
  of fractions, 79
  of quadratic forms, 59, 63
  relation, 64, 79
    reflexivity, 64, 79
    symmetry, 64, 79
    transitivity, 64, 79
Euclid
  algorithm, 3
  ancestor of ACC, 191
  *Elements*, 1, 11, 24, 101
  formula for Pythagorean triples, 35, 36
  induction, 24, 37
  prime divisor property, 1, 189
  proved infinitude of primes, 2

Euclidean algorithm, 1, 3
  and continued fraction, 4
  by division with remainder, 4
  history, 25
  and irrationality, 11
  for polynomials, 16, 31, 84
  by subtraction, 3
  symbolic form, 6
Euclidean domain, 17
Euler, Leonhard
  conjecture on $x^2 + 5y^2$, 60
  phi function, 73
  solution of $y^3 = x^2 + 3$, 50
  on sums of squares, 46
  used algebraic integers, 31, 33
extension field, 89
  algebraic, 90
  basis, 90, 91
  dimension, 90

factor
  of ideal, 104
  of polynomial, 22
  of sum of squares, 45
  theorem, 22
    for multivariable polynomials, 23, 166
    for polynomials over a ring, 22
Fermat, Pierre de
  conjecture on $x^2 + 5y^2$, 60
  and Diophantine equations, 33
  last theorem, 33
  last theorem for $n = 4$, 36
  little theorem, 20, 29, 49
    and binomial coefficients, 29
  and Pell equation, 25
  primes, 24
  on primes of form $x^2 + ky^2$, 59
  proof using Pythagorean triples, 206
  theorems on primes, 60
  two square theorem, 33, 49
    Lagrange proof, 65
Ferrari, Lodovico, 100
Fibonacci
  numbers, 4
  proved Diophantus identity, 39
field, 1, 19, 30
  algebraic number, 30, 86
  of algebraic functions, 184
  axioms, 19, 79
  called Körper by Dedekind, 30
  cyclotomic, 164, 178

extension, 89
　dimension, 131
　finite, 19, 30
　　as quotient, 112
　of fractions, 82
　isomorphism, 87
Fontana, Niccolò, 26
four-group, 69
fraction
　of a domain, 82, 183
　field, 82
fractions, 79
　product, 80
　sum, 80
free module, 171
　basis, 180
　definition, 180
　rank, 178
　　definition, 180
FTA *see* fundamental theorem of algebra, 86
fundamental theorem
　of algebra, 86, 99, 140
　　and calculus, 205
　　provable in $RCA_0$, 125
　of arithmetic, 1, 9
　of finite Abelian groups, 68

Galois, Evariste, xi, 100
　discovered finite fields, 30
　group, xi
Gauss, Carl Friedrich
　composition of forms, 67
　on determinants, 169
　*Disquisitiones*, 62, 76
　lemma, 96
　proofs of FTA, 100
　theorem on primitive root, 70
　theory of quadratic forms, 62
Gaussian elimination, 144
　and computation of det, 155
Gaussian integer, 40
　conjugate, 41
　division property, 42, 44
　norm, 41
Gaussian prime, 41
　classification, 47
　factorization
　　existence, 41
　　uniqueness, 42–44
gcd. *See* greatest common divisor, 3
generator of ideal, 108

generators and relations, 179
generators of module, 172, 179
genus, 206
　and rational functions, 209
　of Riemann surface, 210
　topological interpretation, 209
Gödel, Kurt, 123
Grassmann, Hermann, 25
　on arithmetic, 25, 146
　exchange lemma, 130
　inner product, 145
　linear geometry, 145
　on outer product and det, 170
greatest common divisor, 1, 3, 9
　in Dedekind domain, 201
　of ideals, 190
　of polynomials, 16, 84
　in $\mathbb{Z}[\sqrt{-5}]$, 191
group, xi, 64
　Abelian, 69, 127
　cyclic, 69, 70
　direct product, 69
　finite Abelian, 68
　Galois, xi
　solvable, xii

Halmos, Paul, 148
Hamel, Georg, 129
Hilbert, David, 32
　basis theorem, 116, 117
　"not mathematics but theology,", 117
homomorphism, 104
　canonical, 111
　of groups, 112
　kernel, 111
　preserves structure, 147
　of rings, 111
　of vector spaces, 134
Horner's method, 28

ideal, xii, 68, 71, 104
　class, 174
　conjugate, 191
　definition, 108
　division, 190
　fractional, 197
　gcd, 190
　generators, 108
　inverse, 189, 195
　　in Dedekind domain, 197
　as kernel of homomorphism, 111

ideal (Cont.)
  lcm, 191
  maximal, 112, 113
    existence, 116
    inverse of, 197
    is prime, 192
  as a module, 173
  nonprincipal, 108
  number, 104
    in $\mathbb{Z}[\sqrt{-5}]$, 106
  prime, 104, 189
    definition, 192
    in Noetherian domain, 194
    not maximal, 193
    quotient characterization, 192
    of $\mathbb{Z}_E$, 196
  prime factorization, 78
  principal, 8, 65, 108
    inverse of, 198
  product, 194
  unique prime factorization, 104, 199
    in $\mathbb{Z}$, 108
    in $\mathbb{Z}[\sqrt{-5}]$, 106
induction
  in Dedekind, 25
  in Euclid, 24
  in Grassmann, 25
  in Pascal, 28
  in Peano axioms, 25
  and well-ordering, 24
infinity
  axioms of, 123
  countable, 119
  of prime numbers, 3
  uncountable, 121
inner product, 145
integer, 5
  algebraic, 11, 14, 31, 53
    defined by Dedekind, 31
    definition, 93
    factorization of, 95
    second definition, 96
    of algebraic number field, 32, 78, 93, 94, 174
  cyclotomic, 102, 179
  over a domain, 183, 187
  of a field, 183
  Gaussian, 40
  over a ring, 182
    definition, 183
  quadratic, 61
integral

algebraic function, 184, 188
basis, 149, 168, 175
  definition, 176
  existence, 176
closure, 184
  introduced by Noether, 187
  over ring, 183, 184
  is transitive relation, 186
integral domain. *See* domain, 82
integrally closed, 184
intermediate value theorem, 100
inverse
  additive, 18, 127
  ideal, 189, 195
    in Dedekind domain, 197
  of maximal ideal, 197
  mod a prime polynomial, 84
  multiplicative, 19
  of nonzero ideal, 200
  of principal ideal, 198
irrationality, 10
  discovered by Pythagoreans, 10
  and Euclidean algorithm, 11
  of $\sqrt{2}$, 10
    via continued fraction, 12
    via unique prime factorization, 11
irreducible polynomial, 84
  has no multiple root, 140
isomorphism
  of fields, 87
    into $\mathbb{C}$, 142
  leaves dimension invariant, 134
  of vector spaces, 134

Kaplansky, Irving, 148
kernel, 111
Knobloch, Eberhard, 169
Kronecker, Leopold
  on Abelian groups, 68, 76
  avoided actual infinity, 99
  concept of algebraic number field, 31
  opposed to Cantor, 101
  replacement for FTA, 99
  solution field construction, 101
Krull, Wolfgang, 114
  on Dedekind domain properties, 202
  maximal ideal theorem, 116
Kummer, Ernst Eduard, 31, 75
  cyclotomic integers, 102
  ideal numbers, 75, 104
Kuratowski, Kazimierz, 123

Lagrange, Joseph-Louis
  discriminant, 64, 167
  equivalence of quadratic forms, 59
  lemma, 49
  solved general Pell equation, 25
  theorem on polynomials, 22, 49
Laplace, Pierre-Simon, 169
lcm. *See* least common multiple, 9
least common multiple
  in Dedekind domain, 201
  of ideals, 191
Leibniz, Gottfried Wilhelm, 169
lemma
  exchange, 130
  Gauss, 96
  Lagrange, 49
  Zorn, 115, 116
lexicographic order, 119
linear
  algebra, xii
    basis-free, 148
    history, 144
    for modules, 173
  equations, 127
    and Cramer's rule, 145, 158
    and det, 156
    and the determinant, 145
    Diophantine, 6
    over domain, 157
    and Gaussian elimination, 144
    homogeneous, 157
  independence, 90
    of basis elements, 90
    definition, 128
    determinant criterion, 156
  map, 126, 147
    composition, 136
    definition, 134
    injective, 134, 135
    in Lagrange, 147
    matrix for, 136
    of modules, 173
    surjective, 134, 135
  space, 145

matrix, 5, 136
  for algebraic number, 136
  for basis change, 159
  basis-independent properties, 136
  introduced by Cayley, 148
  inverse, 148
  for linear map, 126, 136

depends on basis, 136
multiplication, 136, 137
for quaternion, 139
sum and product, 148
transpose, 155, 165
maximal ideal, 112
  existence, 114, 116
    and axiom of choice, 123
  inverse of, 197
  is prime, 192
  quotient criterion, 113
  of $\mathbb{Z}[\sqrt{-5}]$, 114
minimal polynomial, 86, 97
  is characteristic, 162
  and conjugates, 143
module, 126, 136, 171
  as Abelian group, 172
  basis, 173
  definition, 172
  free, 171
    definition, 180
    rank, 178
  generators, 172, 179
  quotient, 173
  reason for name, 173
monic polynomial, 31, 93, 97
Muir, Sir Thomas, 169
multiplicative property
  of absolute value, 40
  of conjugate, 13, 41
  of det, 58, 66, 149
    proof, 152
  goes from det to norm, 138, 161
  of isomorphism, 163
  of norm, 39, 57
    via the $\sigma_I$, 163

$\mathbb{N}$, 2
  is well ordered, 2, 24
  a potential infinity, 99
*Nine Chapters*, 144
Noether, Emmy, xii
  and algebraic geometry, 188
  ascending chain condition, 191
  and axiom of choice, 119
  on Dedekind domain properties, 202, 204
  introduced ACC, 114, 116
  theory of integers, 187
  theory of rings, 32
  urged reading of Dedekind, 77
  used dependent choice, 116
Noetherian domain, 194

Noetherian ring, 104, 113
  definition, 114
  equivalent properties, 114
  property of $\mathbb{Z}_E$, 182
nonmeasurability, 123
norm
  and absolute value, 57
  of algebraic integer, 57
  of algebraic number, 161
  of complex number, 39
  in cyclotomic field, 164
  as determinant, 58, 149
  of Gaussian integer, 41
  is basis-independent, 149
  is multiplicative, 39, 57, 161
  as product of conjugates, 57, 58
  relative, 161
  in terms of $\sigma_i$, 163
number
  algebraic, 86, 101
  complex, 39, 55
  ideal, 104
  integer, 5
  irrational, 10
  natural, 2
  rational, 10

Pascal, Blaise
  theorems on binomial coefficients,
    28
  triangle, 28
  used induction, 28
Peano, Giuseppe, 25
  axioms for arithmetic, 25, 124
  axioms for vector spaces, 146
Pell equation, xii, 13, 75
  and continued fractions, 25
  general, 25
  and $\sqrt{2}$, 25
permutation, 152
  even or odd, 153, 169
  transposition, 153
PID. *See* principal ideal domain, 108
polynomial
  characteristic, xii, 149, 159
  content, 96
  cyclotomic, 98
  derivative, 139
  factor, 22
  as integral rational function, 184
  irreducible, 31, 84

minimal, 86, 97
  and conjugates, 143
monic, 31, 93, 97
number of roots, 22
prime, 84
ring, 15, 83
  multivariable, 84
  $R[x]$, 17
  unique prime factorization, 84
prime, 2
  Gaussian, 41
  ideal, 104, 189
    definition, 192
    divisor property, 194
    indivisibility, 201
    in Noetherian domain, 194
    which is not maximal, 193
    quotient characterization, 192
    of $\mathbb{Z}_E$, 196
  in a ring, 15
  natural number, 1
  of form $x^2 + ky^2$, 60
  of $\mathbb{Z}[\sqrt{-2}]$, 48
  polynomial, 84
prime divisor property, 6
  defines prime ideal, 189, 192
  for ideals, 194
  in $\mathbb{Q}[x]$, 16
  in $\mathbb{Z}$, 8, 192
  of natural numbers, 1
prime factorization
  existence, 2
  in $\mathbb{Z}[i]$, 41
  uniqueness, 9
prime polynomial. *See* irreducible polynomial,
    84
primitive element, 92
  theorem, 139
    proof, 140
primitive root, 70
principal ideal, 65, 88
  definition, 108
  domain, 8, 10, 108, 173, 180, 190
  inverse of, 198
  notation, 108, 112
  shape, 174
  of $\mathbb{Z}[\sqrt{-5}]$, 174
product of ideals, 194
Pythagorean
  discovery of irrationality, 10
  equation, 75

theorem, 44, 52
  and inner product, 146
triples, 34, 206
  Euclid's formula, 35
  primitive, 36

Q, 10
Q[x], 15
  has unique prime factorization, 17
quadratic form, 59
  composition, 66
  discriminant, 59
  equivalence, 59, 63
  and quadratic integers, 61
  reduced, 64
quaternion, 139
quotient, 2
  of algebraic integers, 83
  of modules, 173
  of ring by ideal, 71, 88, 104
    definition, 111

ℝ, 70, 99
  completeness of, 101
  has no zero divisors, 82
  is uncountable, 101, 121
  as vector space over Q, 128
rank of free module, 178
  analogous to dimension, 180
  definition, 180
rational
  function, 184
  number, 10
    as equivalence class of fractions, 79
  points
    on circle, 34
    on ellipse, 36
    on hyperbola, 35
RCA₀, 125
reflexivity, 64, 79
remainder, 2
Riemann surface
  branch point, 207
  as covering, 207, 209
  from a function field, 210
  genus, 210
  point of, 210
  sheets, 207
Riemann, Bernhard
  and rational functions, 206
  sphere, 208

Riemann-Hurwitz formula, 210
ring, xii, 1, 30
  of algebraic functions, 188
  of algebraic integers, 32, 93, 94, 103
  axioms, 15, 79
  finite, 21
  homomorphism, 111
  of integers of a field, 183
  integral over another, 184
  integral over subring, 185
  Noetherian, 104, 113
    definition, 114
    equivalent properties, 114
  nonzero, 114
  polynomial, 15, 83
    multivariable, 84
  quotient by ideal, 71, 88, 104
    definition, 111
  R[x], 17
  with zero divisors, 83
  ℤ[i], 40
Rothe, Heinrich August, 169

scalar, 127
Schönemann, Theodor, 97
Seki, Takakazu, 169
Solovay, Robert, 123
solution field, 87
solvability, xii
  of linear Diophantine equation, 7
span, 90, 128
Steinitz, Ernst, 130
Stevin, Simon, 16
straightedge and compass, 133
submodule, 173
  of free ℤ-module, 181
subspace
  definition, 132
  dimension, 132
  of vector space, 126
symmetry
  of equivalence relation, 64, 79
  of rectangle, 69

Tartaglia. See Fontana, Niccolò, 26
theorem
  binomial, 28
  Chinese remainder, 71
  Dedekind product, 131
  factor, 22
  Fermat two square, 33, 49, 65

theorem (Cont.)
  Fermat's last, 33
  Fermat's little, 20, 29, 49
  fundamental
    of algebra, 86, 99, 125, 140
    of arithmetic, 1, 9
    of finite Abelian groups, 68
  Hilbert basis, 116
  intermediate value, 100
  Lagrange
    on polynomials, 22, 49
  primitive element, 139, 140
  Pythagorean, 44, 52
  well-ordering, 123
trace, xii, 149
  of algebraic number, 161
  in cyclotomic field, 164
  definition, 160
  is basis-independent, 136, 149, 161
  is coefficient of charpoly, 160
  relative, 161
  in terms of $\sigma_i$, 163
transitivity
  of equivalence relation, 64, 79
  of relation "integral over," 186
transpose of matrix, 155, 165
transposition, 153

uncountability, 101, 121
  of $\mathbb{R}$, 121
unimodular map, 64
unique prime factorization, xii
  assumed by Euler, 31
  and class number, 65
  fails in $\mathbb{Z}[\sqrt{-5}]$, 61, 105
  holds in PIDs, 109
  of ideals, 104, 189, 199
  Kummer's replacement, 75
  may require ideals, 104
  in $\mathbb{N}$, 1, 9
  for polynomials, 84
  and principal ideals, 65
  in $\mathbb{Q}[x]$, 17
  of sum of squares, 46
  in $\mathbb{Z}$, 9
  in $\mathbb{Z}[i]$, 43, 44
  in $\mathbb{Z}[\sqrt{-2}]$, 45
  in $\mathbb{Z}[\sqrt{2}]$, 19
  in $\mathbb{Z}[\zeta]$, 53
unit, 15
  of $\mathbb{Z}[i]$, 44

of $\mathbb{Z}[\sqrt{-2}]$, 51
of $\mathbb{Z}[\sqrt{2}]$, 15

Vandermonde, Alexandre-Théophile, 166
  founded determinant theory, 169
vector, 127
vector space, 93, 126
  as Abelian group, 127
  axioms, 127, 146
  basis, 126
    and axiom of choice, 123, 128
  concept in Grassmann, 145
  definition, 127
  dimension, 126, 128
    is well-defined, 130
  finite-dimensional, 130
  homomorphism, 134
  isomorphism, 134
Viète, Francois, 27
  angle property, 56
Vitali, Giuseppe, 123

Weber, Heinrich Martin
  algebraic function theory, 188
Weil, André, 60, 75
well-ordering
  and axiom of choice, 120, 123
  definition, 120
  and induction, 24
  of $\mathbb{N}$, 2
  proved by Zermelo, 123
  theorem, 123
  and Zorn's lemma, 120
Wiles, Andrew, 36
$\mathrm{WKL}_0$, 125

$\mathbb{Z}$, 5
  is principal ideal domain, 8, 10
  is ring, 15
  $\mathbb{Z}[i]$, 40
  $\mathbb{Z}_E$, 78, 95
    has integral basis, 176
    is Dedekind domain, 197
    is free $\mathbb{Z}$-module, 186
    is Noetherian, 182
    prime ideal factorization, 104
    prime ideals, 186, 196
Zermelo, Ernst, 123
zero divisor, 17, 21, 80, 82
$\mathbb{Z}_n$, 73
Zorn, Max, 115

Zorn's lemma, 115
    and axiom of choice, 123
    and Hamel basis, 129
    and maximal ideals, 116
    and well-ordering, 120
    statement, 116
$\mathbb{Z}\left[\sqrt{-2}\right]$, 45
$\mathbb{Z}\left[\sqrt{-5}\right]$, 61
    division property fails, 63

gcd in, 105, 107
ideal in, 106
is Noetherian, 118
maximal ideals, 113
nonprincipal ideal, 108, 204
nonunique prime factorization, 61, 105
prime ideal, 192
principal ideals, 174
quotients of, 113

Printed in the United States
by Baker & Taylor Publisher Services

Printed in the United States
by Baker & Taylor Publisher Services